GUIDELINES FOR
**SAFE WAREHOUSING
OF CHEMICALS**

This is a publication of the
CENTER FOR CHEMICAL PROCESS SAFETY
of the
AMERICAN INSTITUTE OF CHEMICAL ENGINEERS
A complete list of CCPS publications can be found at the end of this book

GUIDELINES FOR
SAFE WAREHOUSING
OF CHEMICALS

CENTER FOR CHEMICAL PROCESS SAFETY
of the
American Institute of Chemical Engineers
345 East 47th Street, New York, NY 10017

Copyright © 1998
American Institute of Chemical Engineers
345 East 47th Street
New York, New York 10017

All rights reserved. No part of this publication may be reproduced, stored in a retrieval system, or transmitted in any form or by any means, electronic, mechanical, photocopying, recording, or otherwise without the prior permission of the copyright owner.

Library of Congress Cataloging-in Publication Data
Guidelines for safe warehousing of chemicals.
    p.   cm.
Includes bibliography and index.
ISBN 0-8169-0659-9
  1. Chemicals—Storage—Safety measures. 2. American Institute of Chemical Engineers. Center for Chemical Process Safety. II. Title: safe warehousing of chemicals.
TP201.N85 1998   97-30661
660´.2804—dc21   CIP

This book is available at a special discount when ordered in bulk quantities. For information, contact the Center for Chemical Process Safety at the address shown above.

It is sincerely hoped that the information presented in this volume will lead to an even more impressive safety record for the entire industry; however, the American Institute of Chemical Engineers, its consultants, CCPS Subcommittee members, their employers' officers and directors and Schirmer Engineering Corporation disclaim making or giving any warranties or representations, express or implied, including with respect to fitness, intended purpose, use or merchantability and/or correctness or accuracy of the content of the information presented in this document. As between (1) American Institute of Chemical Engineers, its consultants, CCPS Subcommittee members, their employers, their employers' officers and directors, and Schirmer Engineering Corporation and (2) the user of this document, the user accepts any legal liability or responsibility whatsoever for the consequences of its use or misuse.

Contents

Preface xi
Acknowledgment xiii
Acronyms xv

1
Introduction
1.1. Background 1
1.2. Scope 1
1.3. Purpose 2

2
Commodity Hazards
2.1. Synopsis 3
2.2. Identification of Chemicals 3
2.3. Properties and Hazard Identification of Chemicals 5
2.4. Systems for Commodity Classification 5
    2.4.1. Environmental Protection Agency 11
    2.4.2. National Fire Protection Association 12
    2.4.3. National Paint and Coatings Association's Hazardous Materials Identification System 16
    2.4.4. United Nations (UN) and Department of Transportation (DOT) Hazardous Materials Classes 17
2.5. Container and Packaging Systems 17
2.6. Commodity Compatibility and Separation 18
References 20
Additional Reading 23

3
Administrative Controls

3.1. Synopsis	25
3.2. Safety and Risk Management Policies	25
3.3. Hazard and Risk Management	26
3.4. Control of Ignition Sources	26
3.5. Regulatory Compliance	26
3.6. Risk Management Organization	28
3.7. Employee Hiring, Training and Operations	28
3.7.1. Employee Hiring	*28*
3.7.2. Training	
3.7.3. Operations	*29*
3.8. Housekeeping	29
3.9. Inventory Management	29
3.10. Management of Change	31
References	32
Additional Reading	33

4
Employee Safety and Health

4.1. Synopsis	35
4.2. Policy	35
4.3. Administrative and Engineering Controls	36
4.3.1. Administrative Controls	*36*
4.3.2. Engineering Controls	*37*
4.4. Hazard Communication	37
4.4.1. Labels	*38*
4.4.2. Material Safety Data Sheets	*38*
4.4.3. Employee Information and Training	*39*
4.5. Personal Protective Equipment	40
4.5.1. Implementing a PPE Program	*40*
4.5.2. Selecting PPE	*40*
4.5.3. Chemical Protective Clothing	*41*
4.5.4. Foot Protection	*43*
4.5.5. Head Protection	*43*
4.5.6. Eye and Face Protection	*43*
4.5.7. Hand Protection	*43*
4.5.8. Respirators	*44*

Contents   vii

4.5.9. Respirator Selection	*46*
4.5.10. Respirator Usage	*46*
4.5.11. Training	*46*
4.5.12. Maintenance and Inspection	*48*
4.6. Safety Equipment	48
4.7. Emergency Response Training	49
4.7.1. Emergency Spill Response	*49*
4.7.2. Manual Fire Fighting	*52*
4.7.3. First Aid	*52*
References	52
Additional Reading	53

5
Site Considerations

5.1. Synopsis	55
5.2. Health and Environmental Exposure	55
5.2.1. Baseline Environmental Assessment	*56*
5.2.2. Population Proximity, Density, and Sensitivity	*56*
5.2.3. Warehouse Truck Traffic	*56*
5.2.4. Highly Sensitive Environments	*57*
5.2.5 Surface Water, Ground Water, and Soil Permeability	*57*
5.3. Natural Peril Exposures	57
5.3.1. Earthquake	*58*
5.3.2. Flood	*59*
5.3.3. Hurricanes	*59*
5.3.4. Tornadoes	*60*
5.3.5. Lightning	*62*
5.3.6. Arctic Freeze	*62*
5.4. Exposures from Surrounding Activities	63
5.4.1. Adjacent Facilities, Airports, Highways, and Railroads	*63*
5.4.2. High Pressure Flammable Gas and Liquid Transmission Lines	*63*
5.4.3. Riot and Civil Commotion	*63*
5.5. Emergency Responders	64
5.6. Adequacy and Reliability of Public Utilities	64
References	64
Additional Reading	65

6
Design and Construction

6.1. Synopsis	67
6.2. Construction Documents—Approvals and Permits	67
6.3. Means of Egress	70
6.3.1. Travel Distance	*72*
6.4. Environmental Protection	73
6.4.1. Containment and Drainage Capacity Considerations	*73*
6.4.2. Warehouse Floor System	*82*
6.4.3. Concrete Criteria	*82*
6.4.4. Surface Preparation	*83*
6.4.5. Coatings and Sealers	*84*
6.4.6. Maintenance and Repair of the Floor	*85*
6.4.7. Airborne Effluent	*86*
6.5. Fire Mitigation Construction Features	86
6.5.1. Fire-Rated Separations	*89*
6.5.2. Protection of Openings and Penetrations	*93*
6.5.3. Through-Penetrations	*94*
6.5.4. Heat and Smoke Venting	*97*
6.5.5. Powered Ventilation Systems	*99*
6.5.6. Emergency and Standby Power Systems	*99*
6.6. Deflagration Prevention and Mitigation	99
6.6.1. Temperature Control	*100*
6.6.2. Gas and Vapor Control	*101*
6.6.3. Sources of Ignition	*102*
6.6.4. Spatial Separation	*103*
6.6.5. Damage Limiting Construction	*103*
6.7. Natural Peril Mitigation	104
6.7.1. Earthquake	*105*
6.7.2. Flood	*106*
6.7.3. Lightning	*107*
6.7.4. Windstorm, Hurricane and Tornado	*107*
6.8. Security Features	107
6.9. Outdoor Storage	108
References	109
Additional Reading	111

7
Fire Protection Systems

- 7.1. Synopsis — 113
- 7.2. Storage Considerations — 114
- 7.3. Fire Control, Suppression, and Extinguishing Systems — 115
  - *7.3.1. Fire Control* — *116*
  - *7.3.2. Fire Suppression* — *117*
  - *7.3.3. Fire Extinguishment* — *119*
  - *7.3.4. Fire Extinguishers* — *121*
- 7.4. Fire Detection and Alarm Systems — 121
- References — 126
- Additional Reading — 127

8
Inspection, Testing, and Maintenance Programs

- 8.1. Synopsis — 129
- 8.2. Inspection and Test Programs — 129
  - *8.2.1. Program Objectives* — *129*
  - *8.2.2. Critical Equipment and Construction Features* — *129*
  - *8.2.3. Inspection and Test Program Elements* — *130*
  - *8.2.4. Maintenance* — *131*
  - *8.2.5. Maintenance Procedures* — *132*
- References — 132
- Additional Reading — 134

9
Emergency Planning

- 9.1. Synopsis — 135
- 9.2. Loss Scenarios — 135
- 9.3. Plan Objectives — 135
  - *9.3.1. Employees* — *136*
  - *9.3.2. Surrounding Population* — *136*
  - *9.3.3. Environment* — *136*
  - *9.3.4. Property Protection and Business Interruption* — *136*
- 9.4. Plan Development — 136
- 9.5. Plan Elements — 137

9.5.1. Policy Statement *137*
9.5.2. Scope and Objectives *137*
9.5.3. Pre-Incident Planning *137*
9.5.4. Incident Response *140*
9.6. Emergency Spill Response 141
9.6.1. Planning
9.6.2. Responding to a Hazardous Material Spill *143*
9.6.3. Cleanup *143*
9.6.4. Reporting *144*
9.6.5. Public Response *146*
9.7. Regulations and Resources 146
9.7.1. U.S. Regulations *146*
9.7.2. CMA Responsible Care Program *148*
References 149
Additional Reading. 149

10
Selected Research and Discussion Topics
10.1. Synopsis 151
10.2. Commodity Hazards and Fire Protection Systems 151
10.3. Design and Construction 152

Appendix A
Summary of NFPA 704 Marking System 153

Appendix B
Summary of HMIS 155

Appendix C
United Nations and U.S. Department of Transportation
Hazardous Materials Classes 157

Appendix D
Additional Resources 160

Glossary of Terms *161*
Index *169*

Preface

The American Institute of Chemical Engineers (AIChE) has more than a 30-year history of involvement with process safety for chemical processing plants. Through its strong ties with process designers, builders, operators, safety professionals and academia, the AIChE has enhanced communication and fostered improvement in the high safety standards of the industry. AIChE publications and symposia have become an information resource for the chemical engineering profession on the causes of accidents and means of prevention.

The Center for Chemical Process Safety (CCPS) was established in 1985 by the AIChE to develop and disseminate technical information for use in the prevention of major chemical accidents. The CCPS is supported by a diverse group of industrial sponsors in the chemical process industry and related industries who provide the necessary funding and professional guidance for its project. The CCPS Technical Steering Committee and the individual technical subcommittees overseeing individual projects are staffed by representatives of sponsoring companies. The first CCPS/AIChE project was the preparation of *Guidelines for Hazard Evaluation Procedures.* Since that time a number of Guidelines and shorter Concept Books, have been produced.

Chemicals are ubiquitous in commerce and industry, and warehousing of chemicals and chemical products is an essential part of those activities. While many of the practices and standards applied to safe warehousing of chemicals are similar to those followed for safe warehousing of a wide range of other materials and commodities, chemicals may present unique safety, environmental or property protection challenges. The purpose of these Guidelines is to identify and address those issues unique to chemical warehousing while recognizing the large areas of commonality with other types of warehousing operations. These

Guidelines are intended for use by warehouse operators, architects, designers and others concerned with safe warehousing of chemicals. Most of the content is general and performance-based, extensive references are made to consensus codes and standards for specific technical information..

This book, *Guidelines for Safe Warehousing of Chemicals*, is the result of a project begun in 1993 with the formation of the Subcommittee on Safe Warehousing of Chemicals. The Subcommittee, representing several major chemical companies, selected and worked with a contractor, Schirmer Engineering Corporation, to produce these Guidelines.

Acknowledgments

The American Institute of Chemical Engineers (AIChE) wishes to thank the Center for Chemical Process Safety (CCPS) and those involved in its operation, including its many sponsors whose funding and technical support made this project possible.

Particular thanks are due the members of the CCPS Safe Warehousing of Chemicals Subcommittee who worked with the authors at Schirmer Engineering Corporation to produce this book. Their dedicated efforts, technical contributions and guidance in the preparation of this book are sincerely appreciated.

The members of the Safe Warehousing of Chemicals Subcommittee are:

David Tabar (Chair), *The Sherwin Williams Company*
John Davenport, *Industrial Risk Insurers*
John LeBlanc, *Factory Mutual Research Corporation*
Gary Page, *American Home Products*
Jim Thomas, *McLaren/Hart Environmental Engineering Corporation*
Anthony Torres, *Eastman Kodak Company*

Former Members were:

John Anderson, *DuPont Company*
John F. Murphy, *Dow USA*

Technical Contributors and Reviewers were:

Robert P. Benedetti, *National Fire Protection Association*
William J. Bradford, *Brookfield, CT*
Daniel A. Crowl, *Michigan Technological University*
John A. Hoffmeister, *Lockheed Martin Energy Systems, Inc.*
Peter N. Lodal, *Eastman Chemical Company*
Gerard Opschoor, *TNO Prins Maurits Laboratorium*

Kenneth Mosig, *American International Underwriters*
Adrian L. Sepeda, *Occidental Chemical Corporation*
Scott A. Stookey, *Austin Fire Department*

The following Schirmer Engineering Corporation personnel authored this guideline:

David P. Nugent, Project Manager
Judy Lyn Freeman
Mark Oliszewicz, P.E.

The authors would like to thank Galina Markina, for her preparation of CAD figures. Additionally, the exceptional dedication and skill in manuscript preparation by Kathie Cronin and April Nelson was invaluable

Finally, we wish to express our appreciation to Jack Weaver and Lee Schaller of the CCPS staff for their support and guidance.

Acronyms

ACGIH	American Conference of Governmental Industrial Hygienists
AFFF	Aqueous Film Forming Foam
AIChE	American Institute of Chemical Engineers
ANSI	American National Standards Institute
ASTM	American Society for Testing and Materials
BFE	Base Flood Elevation
BOCA	Building Official and Code Administrators, International
BOD	Biological Oxygen Demand or Biochemical Oxygen Demand
Btu	British Thermal Units
CAS	Chemical Abstracts Service
CCPA	Canadian Chemical Producers Association
CCPS	Center for Chemical Process Safety
CERCLA	Comprehensive Environmental Response, Compensation and Liability Act of 1980
CFR	Code of Federal Regulations
CHRIS	Chemical Hazards Response Information System
CMA	Chemical Manufacturers Association
DOT	Department of Transportation
EPA	Environmental Protection Agency
EPCRA	Emergency Planning and Community Right to Know Act
FEMA	Federal Emergency Management Agency
FFFP	Film Forming Fluoroprotein
FIRM	Flood Insurance Rate Map
FIS	Flood Insurance Study
FM	Factory Mutual
HAZWOPER	Hazardous Waste Operations and Emergency Response
HMIS	Hazardous Materials Identification System

IDLH	Immediately Dangerous to Life and Health
IRI	Industrial Risk Insurers
IUPAC	International Union of Pure and Applied Chemistry
LFL	Lower Flammable Limit
MSDS	Material Safety Data Sheet
NA	North America
NCP	National Contingency Plan
NFPA	National Fire Protection Association
NIOSH	National Institute for Occupational Safety and Health
NPCA	National Paint and Coatings Association
OSHA	Occupational Safety and Health Administration
PEL	Permissible Exposure Limit
ppm	Parts Per Million
PPE	Personal Protective Equipment
RCRA	Resource Conservation and Recovery Act
RQ	Reportable Quantities
RTECS	Registry of Toxic Effects of Chemical Substances
SADT	Self-Accelerating Decomposition Temperature
SARA	Superfund Amendments and Reauthorization Act
SHFA	Special Hazard Flood Area
TLV	Threshold Limit Value
TCLP	Toxicity Characteristic Leaching Procedure
UBC	Uniform Building Code
UFL	Upper Flammable Limit
UL	Underwriters Laboratories
UN	United Nations
US	United States

1 Introduction

1.1. Background

The chemical industry and the chemicals it produces play a major role in advancing the quality of life and improving our standard of living. These improvements, however, have not been without associated risks.

As a society, we have learned to accept and manage the risks to which we are subjected in everyday life. We understand that a completely risk-free environment is unattainable. The chemical warehouse environment is similarly not entirely risk-free. Given the hazards associated with the storage of many chemicals, the potential exists for injuries, illness, environmental damage, property damage, and business interruption.

These occurrences may be initiated by natural disasters, such as earthquake or flood, or by people or equipment. The impact may be felt for many years, and some situations may produce irreversible effects. Therefore, it is paramount that the risks of chemical warehousing be fully understood and appropriate steps taken to prevent or mitigate losses.

1.2. Scope

These Guidelines address the identification of potential hazards associated with the warehouse storage of chemicals in various container and packaging systems. The hazards that are addressed include health effects, environmental pollution, fire and explosion. These guidelines also present means to minimize risk to employees, the surrounding population, the environment, property and business operations. The issues covered in these guidelines are most readily applicable to new warehouses in the planning stage. However, they may also be used by management to assess existing facilities. The types of facilities covered by this guideline include those that are strictly chemical warehouses and those that store other commodities as well. These warehouses can be located on a manufacturing site or be free-standing facilities. It may be appropriate to apply these guidelines to non-company owned or operated warehouses at which company materials are stored (e.g., third party or public warehouses) as a prudent business practice.

These Guidelines do not cover materials in process, rolling stock and transportation issues, filling, dispensing or repackaging of chemicals, or storage of radioactive or detonable material. It also does not address reactivity issues covered in *Guidelines for Safe Storage of Reactive Materials*, published by CCPS.

1.3. Purpose

Safe warehousing of chemicals requires a multi-faceted approach that integrates a wide spectrum of issues. Consistent with its scope, the purpose of these Guidelines is to act as a resource for warehouse operators, designers and others concerned with safe warehousing of chemicals using a performance based approach.

2 | Commodity Hazards

2.1. Synopsis

Chemicals have a wide range of properties and hazards which must be identified and understood if the conditions of "Safe Warehousing" are to be achieved. A complete understanding of the hazards also requires an assessment of the container and packaging systems and storage arrangements, and conditions that might be encountered during the life of the warehouse, such as fire or flood.

This chapter will present the characteristics that determine the overall commodity hazard which is the first step in understanding and limiting the risk. In the chapters that follow, strategies for limiting the risk to employees, the surrounding population, the environment, and property will be presented.

2.2. Identification of Chemicals

Whether the warehouse is in the planning stage or a fully operational facility, chemical identity is usually the first of many considerations in a comprehensive risk assessment.

The hazards of chemicals and commodities can be ranked by various systems of commodity classifications. Chemical identity can be established through the International Union of Pure and Applied Chemistry (IUPAC) name, trade name, common name, United Nations/North America (UN/NA) number, Chemical Abstract Service (CAS) registry number, Registry of Toxic Effects of Chemical Substances (RTECS) number or chemical formula. As an example, acrylonitrile can be identified as follows:

- IUPAC Name: Cyanoethylene
- Trade and Common Name (Synonyms): Acrylon, Acrylonitrile, AN, Carbacryl, Fumigrain, Propenenitrile, VCN, Ventox, Vinyl Cyanide
- (UN/NA) Number: 1093
- CAS Registry Number: 107-13-1
- RTECS AT5250000
- Chemical Formula: $CH_2=CHCN$

Containerized and packaged chemicals should be readily identifiable through the container labels, shipping papers or material safety data sheets (MSDS). In the event that additional information is needed, there

TABLE 2-1
Chemical Identity, Properties, Hazards, Reactivity, and Special Handling Reference List

ACIGIH Threshold Limit Values for Chemical Substances and Physical Agents and Biological Exposure Indices
Bretherick's Handbook of Reactive Chemical Hazards
29 CFR 1910.1000, Toxic and Hazardous Substances
49 CFR Parts 100-179, Hazardous Materials
Condensed Chemical Dictionary (Lewis, 1997)
Farm Chemicals Handbook
Hazardous Chemicals Desk Reference (Lewis, 1993)
The Merck Index (Windzholz)
NFPA 30 Flammable and Combustible Liquids
NFPA 30B Manufacture and Storage of Aerosol Products
NFPA 43B Storage of Organic Peroxide Formulations
NFPA 43D Storage of Pesticides in Portable Containers
NFPA 49 Hazardous Chemicals Data
NFPA 325 Fire Hazard Properties of Flammable Liquids, Gases and Volatile Solids
NFPA 430 Storage of Liquid and Solid Oxidizers
NFPA 490 Storage of Ammonium Nitrate
NFPA 491M Hazardous Chemical Reactions
NIOSH/OSHA Occupational Health Guidelines for Chemical Hazards
NIOSH Pocket Guide to Chemical Hazards
Sax's Dangerous Properties of Industrial Materials
U.S. Dept. of Transportation Emergency Response Guidebook
U.S. Dept. of Transportation/U.S. Coast Guard Chemical Hazard Response Information System (CHRIS)

are a number of useful references, as listed in Table 2-1, where extensive listings of chemicals can be found.

2.3. Properties and Hazard Identification of Chemicals

The physical properties of chemicals determine the type of container and packaging system used and the mode of dispersion if container or package failure occurs. Physical properties also determine the hazards presented. For example, the boiling point of a liquid will affect the rate of vapor evolution if a spill occurs. A low boiling point flammable liquid will increase fire or explosion hazards. Table 2-2 lists some examples of the physical properties of a number of chemicals, including physical state, specific gravity, vapor density, boiling point and water solubility. The fire and explosion hazards of the same chemicals are listed in Table 2-3. The table includes the flash point, auto-ignition temperature, flammable limits, and data on self-reactivity and instability.

Some of the hazards to humans of these example chemicals are listed in Table 2-4. The human health hazards listed include warnings regarding an acute exposure as well as concentrations considered Immediately Dangerous to Life and Health (IDLH), Toxicity by Ingestion (LD_{50}), and Special Hazards of Combustion Products. The OSHA Permissible Exposure Limits (PEL) and ACGIH Threshold Limit Values (TLV) are also listed and are used as an index of chronic health hazard. Additionally, some of the environmental hazards of these chemicals are listed in Table 2-5. The environmental hazards listed include Aquatic Toxicity, Waterfowl Toxicity, Biological Oxygen Demand, and Food Chain Concentration Potential.

2.4. Systems for Commodity Classification

Relative ranking systems have been developed to characterize the degree of hazard associated with a given class of stored commodity. The systems that are pertinent to a chemical warehouse include those that rank aerosol products, compressed and liquefied gases, flammable and combustible liquids, hazardous wastes, ordinary commodities, organic peroxide formulations, pesticides, and liquid and solid oxidizers. In addition to ranking the hazards of the commodities, the rankings are used to establish conditions necessary for prevention and mitigation of unwanted events. Table 2-6 lists some commodity classification references.

2. Commodity Hazards

TABLE 2-2
Physical Properties

Chemical Name	Physical State @ 70°F (21°C) and 1 atm. (1 Bar)	Specific Gravity (H_2O = 1)	Vapor-Air Density (Air = 1)	Boiling Point °F (°C)	Water Soluble
Ammonia	Gas	N/A	.6	−28 (−33)	Yes
Chlorine	Gas	N/A	2.44	−29 (−34)	No
Propane	Gas	N/A	1.6	−44 (−42)	No
Acetone	Liquid	.8	2.0	133 (56)	Yes
Acrylonitrile	Liquid	.8	1.8	171 (77)	Yes
Corn Oil	Liquid	.9	Not Reported	Not Reported	No
Diazinon	Liquid	1.117	Not Reported	Very High; Decomposes	Slight
Ethyl Ether	Liquid	.7	2.6	95 (35)	Slight
Mineral Spirits	Liquid	.8	3.9	300 (149)	No
Motor Oil	Liquid	.84–.96	Not Reported	Very High	No
Ammonium Nitrate	Solid	1.72	N/A	N/A	Yes
Calcium Hypochlorite (Hydrated & Unhydrated)	Solid	2.35	N/A	N/A	Yes (Reacts with water releasing chlorine gas.)
Carbaryl	Solid	1.23	N/A	N/A	No

Sources: NFPA 49, NFPA 325, U.S. Coast Guard CHRIS Manual, and Condensed Chemical DictionaryN/A = Not Applicable

2.4. Systems for Commodity Classification

TABLE 2-3
Fire and Explosion Hazard Data

Chemical Name	Flash Point °F (°C)	Auto-Ignition Temperature °F (°C)	Lower Flammable Limit	Upper Flammable Limit	Self-Reactivity and Instability
Ammonia	N/A	1,204 (651)	15	28	N/A
Chlorine	N/A	N/A	N/A	N/A	N/A
Propane	N/A	842 (450)	2.1	9.5	N/A
Acetone	–4 (–20)	869 (465)	2.5	12.8	N/A
Acrylonitrile	32 (0)	898 (481)	3	17	Hazardous polymerization may be caused by elevated temperature.
Corn Oil	490 (254)	740 (393)	Not Reported	Not Reported	N/A
Diazinon	82–105 (28–41) (solutions only; pure liquid difficult to burn)	N/A	N/A	N/A	N/A
Ethyl Ether	–49 (–45)	356 (180)	1.9	36	May form explosive peroxides.
Mineral Spirits	104 (40)	473 (245)	.8 @ 212°F (100°C)	Not Reported	N/A
Motor Oil	275–600 (135–316)	325-625 (163-329)	Not Reported	Not Reported	N/A
Ammonium Nitrate	N/A	N/A	N/A	N/A	Decomposes above 410°F (210°C) releasing nitrous oxide.
Calcium Hypochlorite (Hydrated & Unhydrated)	N/A	N/A	N/A	N/A	Decomposes above 350°F (177°C) releasing oxygen and chlorine.
Carbaryl	N/A	N/A	N/A	N/A	N/A
Sources: NFPA 49 and NFPA 325			N/A—Not Applicable		

TABLE 2-4
Human Health Hazards

Chemical Name	Acute Health Hazards	IDLH and Toxicity by Ingestion	Special Hazards of Combustion Products	Chronic Health Hazards
Ammonia	VAPOR Poisonous if inhaled. Irritating to eyes, nose and throat. LIQUID Will burn skin and eyes. Harmful if swallowed. Will cause frostbite.	500 ppm Not Pertinent	Not Reported	OSHA PEL: 50 ppm ACGIH TLV: 25 ppm
Chlorine	VAPOR Poisonous if inhaled. Will burn eyes. LIQUID Will burn skin and eyes. Will cause frostbite.	25 ppm Not Pertinent	Toxic products are generated when combustibles burn in chlorine.	OSHA PEL: 1 ppm ACGIH TLV: 1 ppm
Propane	VAPOR Not irritating to eyes, nose or throat. If inhaled, will cause dizziness, difficult breathing, or loss of consciousness. LIQUID May cause frostbite.	20,000 ppm Not Pertinent	Not Reported	OSHA PEL: 1,000 ppm ACGIH TLV: Asphyxiant
Acetone	VAPOR Irritating to eyes, nose and throat. If inhaled, may cause difficult breathing or loss of consciousness. LIQUID Irritating to eyes, not irritating to skin.	20,000 ppm LD_{50} = 5 to 15 g/kg (dog)	Not Reported	OSHA PEL: 1,000 ppm ACGIH TLV: 750 ppm
Acrylonitrile	VAPOR Poisonous if inhaled. Irritating to eyes. LIQUID Poisonous if swallowed. Irritating to skin and eyes.	4 ppm LD_{50} = 50 to 500 mg/kg (rat, guinea pig)	When heated or burned, Acrylonitrile may evolve toxic hydrogen cyanide gas and oxides of nitrogen.	OSHA PEL: 2 ppm ACGIH TLV: 2 ppm

Sources: U.S. Coast Guard CHRIS Manual, 29 CFR 1910.1000 and 29 CFR 1910.1045, and ACGIH TLV Indices

2.4. Systems for Commodity Classification

Chemical Name	Acute Health Hazards	IDLH and Toxicity by Ingestion	Special Hazards of Combustion Products	Chronic Health Hazards
Corn Oil	Not harmful.	Data not available. None	Not Reported	OSHA PEL: Not Reported ACGIH TLV: Not Reported
Ethyl Ether	VAPOR Irritating to eyes, nose and throat. If inhaled, will cause nausea, vomiting, headache, or loss of consciousness. LIQUID Irritating to skin. Harmful if swallowed.	19,000 ppm LD_{50} = 0.5 to 5 g/kg	Not Reported	OSHA PEL: 400 ppm ACGIH TLV: 400 ppm
Mineral Spirits	LIQUID Irritating to skin and eyes. Harmful if swallowed.	Data not available. LD_{50} = 0.5 to 5 g/kg	Not Reported	OSHA PEL: 500 ppm ACGIH TLV: Data not available.
Motor Oil	LIQUID Irritating to skin and eyes. Harmful if swallowed.	Data not available. LD_{50} = 5 to 15 g/kg	Not Reported	OSHA PEL: Not Reported ACGIH TLV: Data not available.
Diazinon	LIQUID Poisonous if swallowed. Irritating to skin and eyes.	Data not available. Grade 3; oral LD_{50} = 76 mg/kg (rat)	Oxides of sulfur and of phosphorus are generated in fires.	OSHA PEL: Not Reported ACGIH TLV: 0.1 mg/m³
Ammonium Nitrate	DUST Irritating to eyes, nose and throat. If inhaled, may cause coughing or difficult breathing.	Not Pertinent Data not available.	Decomposes, giving off extremely toxic oxides of nitrogen.	OSHA PEL: Not Reported ACGIH TLV: Not pertinent.
Calcium Hypochlorite	SOLID Irritating to skin and eyes. If swallowed, will cause nausea, vomiting or loss of consciousness.	Data not available. LD_{50} above 15 g/kg	Poisonous gases may be produced when heated.	OSHA PEL: Not Reported ACGIH TLV: Not Pertinent.
Carbaryl	SOLID or SOLUTION Irritating to skin and eyes. Harmful if swallowed.	625 mg/m³ Grade 2; LD_{50} = 0.5 to 5 g/kg (rat LD_{50} 0.51 g/kg)	Not Reported	OSHA PEL: Not Reported ACGIH TLV: 5 mg/m³

Sources: U.S. Coast Guard CHRIS Manual, 29 CFR 1910.1000 and 29 CFR 1910.1045, and ACGIH TLV Indices

TABLE 2-5 *Environmental Hazards*

Chemical Name	Aquatic Toxicity					Waterfowl Toxicity	Biological Oxygen Demand	Food Chain Concentration Potential
	Concentration	Time of Exposure	Aquatic Species	Effects	Water Type			
Ammonia	2.0–2.5 ppm 60–80 ppm 8.2 ppm	1-4 days 3 days 96 hr.	Goldfish and yellow perch Crayfish Fathead minnow	not stated LC_{100} TL_M	not stated	120 ppm	Not Pertinent	None
Chlorine	0.08 ppm 10 ppm	168 hr. 1 hr.	Trout Tunicates	TL_M killed	fresh salt	DNA*	None	None
Propane	None	None	None	None	None	None	None	None
Acetone	14,250 ppm 13,000 ppm	24 hr. 48 hr.	Sunfish Mosquito fish	killed TL_M	tap turbid	Not Pertinent	(Theor) 122%, 5 days	None noted
Acrylonitrile	100 ppm 0.05–1 ppm	24 hr. 24 hr.	All fish Bluegill	100% killed lethal	fresh salt	Not Pertinent	70%, 5 days	None noted
Corn Oil	DNA*	DNA*	DNA*	DNA*	DNA*	DNA	DNA*	None
Diazinon	0.025 ppm	96 hr.	Stonefly nymph	TL_M	fresh	LD_{50} = 3.54 mg/kg LD_{50} = 5 days, 90 ppm mallard LD_{50} = 7 days, 68 ppm quail	DNA	DNA*
Ethyl Ether	DNA	DNA	DNA	DNA	DNA	DNA	3%, 5 days	None
Mineral Spirits	DNA	DNA	DNA	DNA	DNA	DNA	8%, 5 days	None
Motor Oil	DNA	DNA	DNA	DNA	DNA	DNA	DNA	None
Ammonium Nitrate	DNA	DNA	DNA	DNA	DNA	DNA	DNA	None
Calcium Hypochlorite	0.5 ppm	Time not specified	Trout	killed	fresh	DNA	Not Pertinent	Not Pertinent
Carbaryl	5.5 ppm 0.013 ppm	96 hr. 48 hr.	Bluegill White shrimp	TL_M TL_M	fresh salt	LD_{50} = 2179 mg/kg	DNA	None observed

Source: U.S. Coast Guard CHRIS Manual. *DNA—Data Not Available

2.4. Systems for Commodity Classification

TABLE 2-6
Commodity Classification Reference List

40 CFR Part 152 Pesticide Products
40 CFR Part 261 Hazardous Wastes
NFPA 30B Manufacture and Storage of Aerosol Products
NFPA 30 Flammable and Combustible Liquids Code
NFPA 231 General Storage (Ordinary Commodities)
NFPA 231C Rack Storage of Materials (Ordinary Commodities)
NFPA 43B Storage of Organic Peroxide Formulations
NFPA 430 Storage of Liquid and Solid Oxidizers
NFPA 55 Compressed and Liquefied Gases
NFPA 704 Standard System for the Identification of the Fire Hazards of Materials
NPCA Hazardous Materials Identification System

It should be recognized that definitions for certain hazardous materials, such as "flammable liquids." may not be the same for different classification systems, such as NFPA and DOT. Therefore, it is important to state which classification system a definition is based upon.

2.4.1. Environmental Protection Agency

2.4.1.1. Pesticides

Pesticides can be designated as Toxicity Category I, II, III or IV by the classification system of the Environmental Protection Agency (EPA) in 40 Code of Federal Regulations (CFR) Part 152 Pesticide Products. Toxicity Category I pesticides are the most hazardous and Toxicity Category IV the least hazardous. This system accounts for acute short-term toxicity and does not account for chronic long term toxicity.

Additionally, this system does not account for fire hazard, reactivity hazard or environmental hazard. The toxicity category of a pesticide is determined from the oral LD_{50}, LC_{50}, Dermal LD_{50} as well as local effects to both skin and eyes.

In addition to the toxicity category designation, pesticides can also be labeled with signal words. Toxicity Category I pesticides are labeled "Danger" or "Poison" depending on the effect of an exposure. Toxicity Category II pesticides are labeled "Warning" and both Toxicity Category III and IV are labeled "Caution." This system has limitations in that the LD_{50} and LC_{50} values are based on exposure to laboratory animals and may not correlate directly to humans. Also, the terms used to describe skin and eye effects are subject to interpretation.

2.4.1.2. Hazardous Wastes

Wastes are considered hazardous according to the Resource Conservation and Recovery Act (RCRA) when they meet one or more criteria including ignitability, corrosivity, reactivity or toxicity as per EPA definitions (see Glossary of Terms). Additionally, wastes are also considered hazardous when they are included in one of the following lists:

- F-List (see 40 CFR Part 261.31, Hazardous Wastes from Nonspecific Sources)
- K-List (see 40 CFR Part 261.32, Hazardous Wastes from Specific Sources)
- P-List (see 40 CFR Part 261.33(e), Discarded Commercial Chemical Products, Off-specification Species, Container Residues and Spill Residues Thereof)
- U-List (see 40 CFR Part 261.33(f), Discarded Commercial Chemical Products, Off-specification Species, Container Residues and Spill Residues Thereof)

2.4.2. National Fire Protection Association

2.4.2.1. Aerosol Products

Aerosol products are designated as Level 1, 2, or 3 by the classification system of NFPA 30B, "Code for the Manufacture and Storage of Aerosol Products." Level 1 aerosols are considered the least hazardous and Level 3 the most hazardous. This system is based upon the fire hazard of these products. Other hazards such as the acute and chronic health hazards, as well as the environmental hazards of aerosol products are not addressed by the NFPA 30B classification system. The fire hazard associated with aerosol products is determined by either a 12-pallet aerosol classification test or the chemical heat of combustion of all of the constituents within the aerosol can.

The 12-pallet classification test results consider the number of opened sprinklers, maximum steel beam temperature at the ceiling, maximum plume temperature and velocity, maximum heat flux, maximum weight loss rate, and net percentage weight loss. The classification of a given aerosol product is based upon suppression or control of the fire and the number of sprinklers that opened during the fire test. An aerosol product is considered a "Level 1" if the fire was well controlled or suppressed, a "Level 2" if the fire was well to marginally well controlled or a "Level 3" if the fire was not well controlled.

2.4. Systems for Commodity Classification

The other method of classifying an aerosol product is by determining the chemical heat of combustion of all of the constituents within the aerosol can. This method provides consistent correlation with the 12-pallet aerosol classification test. The chemical heat of combustion is the product of the theoretical heat of combustion and combustion efficiency. An aerosol product is considered "Level 1" if the chemical heat of combustion is greater than 0 and less than or equal to 8600 Btu/lb (20 KJ/g), "Level 2" if it is greater than 8600 Btu/lb (20 KJ/g) and less than or equal to 13,000 Btu/lb (30 KJ/g), and "Level 3" if it is greater than 13,000 Btu/lb (30 KJ/g). NFPA 30B requires that the classification of an aerosol product be stated on the carton to allow easy identification.

A listing of the theoretical heats of combustion, combustion efficiencies and chemical heats of combustion for a number of chemicals commonly found in aerosol products is shown in Table 2-7.

The NFPA 30B classification system is used to specify or determine the need for certain requirements for building ceiling height, allowable quantities, storage arrangement, automatic sprinkler protection, and building construction features.

2.4.2.2. Flammable and Combustible Liquids

Flammable and combustible liquids can be designated as Class IA, IB, IC, II, IIIA, or IIIB by the classification system of NFPA 30, "Flammable and Combustible Liquids Code." Class IA liquids are considered the most hazardous and Class IIIB the least hazardous. This classification system is based upon the closed-cup flash point temperature and with Class IA and Class IB liquids also the boiling point temperature of the liquid. Liquids are considered flammable if their flash points are below 100°F (37.8°C) and combustible if their flash points are at or above 100°F (37.8°C).

TABLE 2-7
Heats of Combustion

Material	Theoretical Heat of Combustion BTU/lb (kJ/g)	Efficiency	Chemical Heat of Combustion BTU/lb (kJ/g)
Acetone	12,296 (28.6)	0.97	11,909 (27.7)
Corn Oil	15,821 (36.8)	0.96	15,176 (35.3)
Mineral Spirits	19,260 (44.8)	0.92	17,712 (41.2)
Source: Newman, 1994			

Fire testing used to develop automatic sprinkler system design criteria has shown that other factors are important to determine the overall fire hazard of containerized liquids. These factors include the liquid properties, such as liquid burning rate, fire point, specific gravity, water solubility, and viscosity, as well as the container design and size. These important factors have been incorporated into the sprinkler system design tables in Section 4-8 of the 1996 edition of NFPA 30.

2.4.2.3. Ordinary Commodities
Ordinary commodities can be designated as either non-combustible or as combustible Class I, II, III, IV, or as Group A, B, or C plastics by the classification system established by NFPA 231 "Standard for General Storage" and NFPA 231C, "Standard for Rack Storage of Materials." This system is based upon the fire behavior of ordinary commodities stored under automatic sprinklers. Non-combustible commodities are considered the least hazardous and Group A plastics the most hazardous by this system. The relative hazard of a commodity is a function of the combustion properties of the material as well as its configuration in the storage array. There are essentially four methods that are used for classifying ordinary commodities. These include small-scale fire tests, subjective physical comparison, intermediate or full-scale fire tests, and fire tests based upon calorimetry.

The classification system used in NFPA 231 and NFPA 231C is used to specify or determine the need for certain requirements that can prevent or mitigate the effects of fire. These includes automatic sprinkler protection, storage arrangement, and building construction features.

2.4.2.4. Organic Peroxide Formulations
Organic Peroxide formulations are designated as Class I, II, III, IV, or V by NFPA 43B, "Code for the Storage of Organic Peroxide Formulations." Class I organic peroxide formulations are considered the most hazardous and Class V organic peroxide formulations are the least hazardous by this system. The fire hazard and self-reactivity behavior of organic peroxide formulations, under storage conditions when subjected to fire exposure, are the basis of this classification system. Specifically, this system applies to certain commercial formulations that are packaged in containers approved by the U.S. Department of Transportation or the Canadian Ministry of Transport. Organic peroxide formulations that can undergo detonation when exposed to a fire are not addressed by this system. The acute and chronic health hazards as well as the environmental hazards of organic peroxides are also not addressed.

The NFPA 43B classification system addresses organic peroxide formulations that will not sustain combustion through those that burn very

2.4. Systems for Commodity Classification

rapidly. Additionally, the self-reactivity behavior of organic peroxide formulations includes those that have no self-reactivity through those that can undergo spontaneous decomposition resulting in a deflagration.

The definitions applied to organic peroxide formulation behavior can be subject to interpretation. Therefore, in an effort to better identify commercial organic peroxide formulations by class, a listing of typical formulations and their classes can be found in the Appendix of NFPA 43B. This listing includes the name and concentration of the organic peroxide, type of diluent, maximum individual container size and the NFPA 704 hazard identification rating.

The NFPA 43B classification system is used to specify or determine the need for certain requirements that can prevent or mitigate the effects of fire and explosion. These requirements include allowable quantities, automatic sprinkler protection, the need to store certain organic peroxide formulations in a refrigerator or freezer, storage arrangement, and building construction features.

2.4.2.5. Liquid and Solid Oxidizers

Liquid and solid oxidizers are designated as Class 1, 2, 3, or 4 by NFPA 430, "Code for the Storage of Liquid and Solid Oxidizers." Class 1 oxidizers are considered the least hazardous and Class 4 the most hazardous. The behavior of commercially available strengths of liquid and solid oxidizers under storage conditions is the basis for this classification system. This behavior includes fire hazard and reactivity. The acute and chronic health hazards as well as the environmental hazards of liquid and solid oxidizers are not addressed.

The NFPA 430 classification system, the fire hazard behavior of liquid and solid oxidizers that can increase the burning rate or cause spontaneous ignition of a combustible material. Also, the reactivity behavior of liquid and solid oxidizers includes those that can undergo a self-sustained decomposition or an explosive reaction. The definitions used to define liquid and solid oxidizer classes can be subject to interpretation. The NFPA 430 classification system "is based on the Technical Committee's evaluation of available scientific and technical data, actual experience, and it's considered opinion." A listing of typical liquid and solid oxidizers by class can be found in the Appendix of NFPA 430. The oxidizers listed are assumed to be pure materials unless otherwise specified.

The NFPA 430 classification system is used to specify or determine the need for certain requirements that can prevent or mitigate the effects of fire and explosion. These requirements include allowable quantities, automatic sprinkler protection, storage arrangement features, and building construction type.

2.4.2.6. Compressed and Liquefied Gases

Compressed and liquefied gases are classified according to their primary hazard using the definitions in NFPA 55, "Storage, Use, and Handling of Compressed and Liquefied Gases in Portable Cylinders." NFPA 55 classifies compressed and liquefied gases as toxic, pyrophoric, flammable, nonflammable, or oxidizing. The definitions for these materials can be found in the Glossary of Terms.

2.4.2.7 NFPA 704, "Identification of the Fire Hazards of Materials"

This system of hazard identification uses a numerical designation of 0 through 4 to provide a relative ranking of health, flammability and reactivity for emergency responders. Additionally, other hazards such as oxidizing potential and water reactivity are also shown. The NFPA 704 markings are diamond shaped and use four color-coded quadrants. Appendix A provides a basic description of the hazard associated with each number. This system provides a general indicator of the type and severity of the hazards associated with a material under fire exposure or spill conditions. The effects of chronic long term health exposure, container and packaging system performance, environmental hazards, and incompatibility with other chemicals are not addressed. **The NFPA 704 fire and reactivity hazard ratings do not correlate with other commodity classification systems such as used in NFPA 430 and 43B.**

2.4.3. National Paint and Coatings Association's Hazardous Materials Identification System

This system of hazard identification is very similar to the NFPA 704 system in that it also uses a relative ranking scale from 0 through 4 for health, flammability and reactivity hazards. The system scaling is the same and commonly uses NFPA 704 fire and reactivity ratings. Terminology used to describe each hazard, however, can differ. Additionally, this system also incorporates what is known as a Personal Protective Index. The Personal Protective Index uses an alphabetical designation and symbols that correspond to a specific combination of protective equipment such as gloves and safety glasses. Appendix B provides a summary of the ratings and a listing of the equipment specified under the Personal Protective alphabetical designation. This system of hazard identification is intended to protect employees against occupational exposures under

2.5. Container and Packaging Systems

conditions of "normal handling." The recommended protective equipment may not be adequate under adverse or upset conditions.

2.4.4. United Nations (UN) and Department of Transportation (DOT) Hazardous Materials Classes

United Nations Committee on Transport of Dangerous Goods and the U.S. Department of Transportation use a classification system, outlined in Appendix C that does not completely correlate with the systems previously described. Furthermore, this system applies to the transportation of commodities and was not intended to be applied to the warehouse environment. It is intended to give an at-a-glance identification of the hazards much like the NFPA 704 and the NPCA Hazardous Materials Identification System.

2.5. Container and Packaging Systems

Container and packaging systems utilize various designs and materials of construction (see Table 2-8). Warehoused chemicals can also be stored in loose bulk piles. The overall hazard of a given commodity is derived from the inherent physical properties, fire and explosion hazards and health

TABLE 2-8
Typical Container and Packaging Systems

	Type	Material
NON-BULK	Bags	Cloth, paper, plastic and paper/foil/plastic composites
	Bag-in-a-box	Plastic bag in corrugated cardboard and polyvinyl alcohol film (water soluble) in corrugated cardboard.
	Bottles	Glass, plastic and polyvinyl alcohol (water soluble).
	Pails	Steel, plastic and fiber.
	Drums	Steel, plastic and fiber.
BULK	Intermediate Bulk Container (rigid, semi-rigid, or flexible)	Steel, aluminum, plastic, wood, fiber, textile and paper.
	Portable Tank	Steel and aluminum

hazards of a particular chemical and the performance of the container or packaging system, if used. Ideally, container and packaging systems should maintain their integrity and not release their contents. In order to fully assess the risk associated with the storage of a particular chemical, all of the conditions or possible scenarios that might occur in a warehouse should be considered. This includes the potential for material handling accidents, water exposure and fire exposure.

The United Nations (UN) has developed criteria for the testing of chemical container and packaging systems. This criteria is "performance-based" and is covered in their publication entitled "Recommendations on the Transport of Dangerous Goods." The U.S. Department of Transportation adopted United Nations recommendations in 1990 and amended them in 1996.

Depending on the container type, UN/DOT "performance-based" testing includes drop, leakproof, hydrostatic pressure, stacking, vibration, top and bottom lift, tear, topple, and righting tests.

Based upon the UN or DOT Hazardous Materials Class, discussed in Section 2.4, a given chemical is also assigned to a "packing group."

There are three "packing groups" as follows:

- Packing Group I—assigned to materials presenting great danger
- Packing Group II—assigned to materials presenting medium danger
- Packing Group III—assigned to materials presenting minor danger

Each "Packing Group" requires that a container or package meet and be maintained to certain performance criteria for the tests previously mentioned. None of these tests evaluates fire performance of containers and packaging systems and therefore will not provide a complete assessment of all of the hazards associated with warehouse storage.

As discussed in Section 2.4, there are four basic methods for assessing the fire hazard of commodities for warehouse storage. This includes small-scale fire tests, subjective physical comparison, intermediate or full-scale fire tests, and fire tests based upon calorimetry. The most accurate assessment of the fire hazard of a commodity will be obtained with intermediate or full-scale fire tests and with some commodities, fire tests based upon calorimetry.

2.6. Commodity Compatibility and Separation

Chemicals are normally packaged in containers that are physically and chemically compatible with them. The release of and accidental contact

2.6. Commodity Compatibility and Separation

between incompatible chemicals can result in a hazardous situation. Incompatible combinations can also occur between the released chemical and containers, packaging materials, air, fire-fighting water, other fire-extinguishing agents, spill clean-up materials, and building components. The accidental contact can be the result of improper material handling, faulty containers or an event such as a fire or earthquake. Chemicals are considered incompatible when their contact can produce a hazardous chemical reaction. This includes exothermic reaction and production of toxic or flammable products. An exothermic chemical reaction can release energy at a rate or quantity that results in a fire or explosion. Additionally, the reaction between incompatible materials might not occur immediately. Some combinations can result in products that are heat, shock, or friction sensitive and subject to reacting later, such as during a clean-up operation.

Table 2-9 lists a number of binary chemical combinations that are chemically incompatible and indicates the results of their coming in contact with each other. Additional information can be obtained by consulting the material safety data sheet (MSDS), the manufacturer or supplier, Bretherick's Handbook of Reactive Chemical Hazards, NFPA 49, NFPA 491M and U.S. Coast Guard CHRIS (Chemical Hazards Response Information System) Manual. Although the cited literature contains an exten-

TABLE 2-9
Examples of Incompatible Binary Combinations

Combinations of A and B		Produce
A	B	
Hydrochloric Acid	Calcium Hypochlorite	Chlorine Gas Release (Toxic)
Primary Amines	Calcium Hypochlorite	n-Chloroamines (Explosive)
Organic Liquids	Calcium Hypochlorite	Violent Reaction
Sodium Cyanide	Sulfuric Acid	Hydrogen Cyanide (Highly Toxic)
Acrylonitrile	Sulfuric Acid	Vigorous Exothermic Reaction
Acetyl Chloride	Water	Violent Reaction
Acetic Anhydride	Water	Violent Reaction
Aluminum Phosphide	Water	Phosphine Gas Release (Pyrophoric and Toxic)
Source: NFPA 491M and NFPA 49		

sive amount of information, the absence of published reactivity data should not infer that no hazard exists.

Incompatible chemicals should be separated by arranging them in segregated, cut-off, or detached storage areas. "Segregated storage" separates the incompatible materials spatially, intervening inert or mutually compatible materials, constructing line of sight barriers, or storing the incompatible material in a fire-resistive cabinet. This approach may be appropriate for materials that are mildly incompatible such as ordinary combustibles and oxidizers that cause an increase in burning rate. Curbing, ramps, or depressed floors can be utilized to restrict the flow of liquids across the floor or to adjacent building areas.

"Cut-off storage" involves partitioning the building into separate compartments. Organic peroxides that present a severe burning rate and reactivity hazard are candidates for cut-off storage. In addition to fire resistance, these walls might also need to be blast resistant.

"Detached storage" involves storage in separate buildings. This storage arrangement is usually reserved for materials that present severe fire, reactivity or health risk. For example, the storage of aluminum phosphide, which reacts with water to produce toxic and pyrophoric phosphine gas, in an unsprinklered detached building will reduce the risk of exposure to the facility and employees, and possibly the surrounding community.

Figure 2-1 illustrates warehouse arrangements incorporating segregated, cut-off, and detached storage.

References

American Conference of Governmental Industrial Hygienists, "Threshold Limit Values for Chemical Substances and Physical Agents and Biological Exposure Indices," Cincinnati, OH, 1996–1997.

Bretherick, L., *Handbook of Reactive Chemical Hazards*, Fifth Edition, Butterworths, London, 1995.

Farm Chemicals Handbook, Volume 82. Meister Publishing Company, Willoughby, Ohio, 1996.

Lewis, R.J., *Condensed Chemical Dictionary*, 13th ed. Van Nostrand Reinhold, New York, 1997.

Lewis, R. J., *Hazardous Chemical Desk Reference*, 3rd ed, Van Nostrand Reinhold, New York, 1993.

National Fire Protection Association. "Flammable and Combustible Liquids Code," NFPA 30, Quincy, MA, 1996.

National Fire Protection Association. "Code for the Manufacture and Storage of Aerosol Products." NFPA 30B, Quincy, MA, 1994.

National Fire Protection Association. "Code for the Storage of Organic Peroxide Formulations." NFPA 43B, Quincy, MA, 1993.

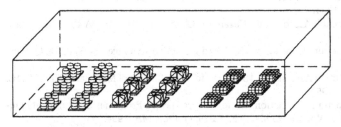

SEGREGATED STORAGE
STORAGE SEPARATED BY DISTANCE
OR INERT MATERIAL

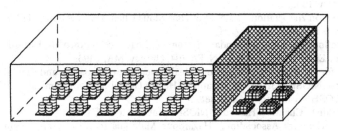

CUT–OFF STORAGE
STORAGE SEPARATED BY A WALL

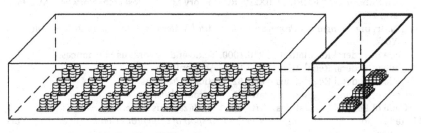

DETACHED STORAGE
STORAGE IN SEPARATE BULIDINGS

LEGEND

- MATERIAL "A"
- MATERIAL "B"
- MATERIAL "C"

NOTE: Materials "A" and "B" are incompatible with each other. Material "C" is inert.

FIGURE 2-1 *Warehouse arrangements incorporating segregated, cut-off, and detached storage.*

2. Commodity Hazards

National Fire Protection Association. "Code for the Storage of Pesticides." NFPA 43D, Quincy, MA, 1994.
National Fire Protection Association. "Hazardous Chemicals Data." NFPA 49, Quincy, MA, 1994.
National Fire Protection Association. "Standard for General Storage." NFPA 231, Quincy, MA, 1995.
National Fire Protection Association. "Standard for Rack Storage of Materials." NFPA 231C, Quincy, MA, 1995.
National Fire Protection Association. "Guide to Fire Hazard Properties of Flammable Liquids, Gases, and Volatile Solids." NFPA 325, Quincy, MA, 1994.
National Fire Protection Association. "Code for the Storage of Liquid and Solid Oxidizers." NFPA 430, Quincy, MA, 1995.
National Fire Protection Association. "Code for the Storage of Ammonium Nitrate." NFPA 490, Quincy, MA, 1993.
National Fire Protection Association. "Manual of Hazardous Chemical Reactions." NFPA 491M, Quincy, MA, 1991.
National Fire Protection Association. "Standard System for the Identification of the Hazards of Materials for Emergency Response." NFPA 704, Quincy, MA, 1996.
National Institute for Occupational Safety and Health/Occupational Safety and Health Administration (NIOSH/OSHA), "Occupational Health Guidelines for Chemical Hazards." NIOSH/OSHA, Cincinnati, OH, 1988.
NIOSH. "Pocket Guide to Chemical Hazards." NIOSH, Cincinnati, OH, 1990.
National Paint and Coatings Association. "Hazardous Materials Identification System Revised—Rating Procedure." Washington, DC, January 1985.
Newman, J.S. "Determination of Theoretical and Chemical Heats of Combustion of Chemical Substances." FMRC J.I. 0X2R8.RC, Factory Mutual Research, Norwood, MA, January, 1994.
Sax, N. Irving. *Dangerous Properties of Industrial Materials*. Van Nostrand Reinhold, New York, 1994.
29 Code of Federal Regulations, 1910.1000, Toxic and Hazardous Substances.
29 Code of Federal Regulations, 1910.1045, Acrylonitrile.
40 Code of Federal Regulations, Part 152, Pesticide Products.
40 Code of Federal Regulations, Part 261, Hazardous Wastes.
49 Code of Federal Regulations, Parts 100-179, Hazardous Materials.
United Nations, *Recommendations on the Transport of Dangerous Goods*, 7th ed. New York, 1991.
U.S. Department of Transportation, "Emergency Response Guidebook." Research and Special Programs Administration, Washington, DC, 1993.
U.S. Department of Transportation, United State Coast Guard, Chemical Hazards Response Information System (CHRIS). *Condensed Guide to Chemical Hazards*, Vol. I. Washington, DC, 1992.
U.S. Department of Transportation, United State Coast Guard, Chemical Hazards Response Information System (CHRIS). *Hazardous Chemical Data Manual*, Vol. II." Washington, DC, 1992.
Windzholz, M. (Ed.). *The Merck Index*, 11th edition, Merck and Co., Inc., Rahway, NJ, 1989.

Additional Reading

Association of American Railroads Bureau of Explosives. *Emergency Handling of Hazardous Materials in Surface Transportation.* Washington, DC, 1992.
Bradford, W.J., "Chemicals." Section 5/Chapter 6, NFPA Handbook, National Fire Protection Association. Quincy, MA, July, 1991.
Bradford, W.J., "Storage and Handling of Chemicals." Section 2/Chapter 27, NFPA Fire Protection Handbook, National Fire Protection Association. Seventeenth Edition, Quincy, MA, July, 1991.
Factory Mutual Engineering Corporation, "Advances in Commodity Classification." Record, Norwood, MA, May/June 1989.
Factory Mutual Engineering Corporation, "Commodities and Storage Arrangements." Record, Norwood, MA, Mar./Apr. 1990.
Factory Mutual Engineering Corporation, "Commodity Classification." Loss Prevention Data 8-0S, Norwood, MA, March, 1991.
Factory Mutual Engineering Corporation, "Flammable Liquids in Drums and Smaller Containers." Loss Prevention Data 7-29, Norwood, MA, September, 1989.
Factory Mutual Engineering Corporation, "Storage of Class 1, 2, 3, 4 and Plastic Commodities." Loss Prevention Data 8-9, Norwood, MA, September, 1993.
Factory Mutual Engineering Corporation, "Storage of Aerosol Products." Loss Prevention Data 7-29S, Norwood, MA, May, 1983.
Fluer, L., *Hazardous Materials Classification Guide.* International Fire Code Institute, Whittier, CA, 1993.
Industrial Risk Insurers, "Commodity Classification Guide." IM.10.0.1, Hartford, CT, June 1, 1992.
Industrial Risk Insurers, "Flammable and Combustible Liquids Storage Facilities." IM.10.2.4, Hartford, CT, December 1, 1992.
McCloskey, C.M., "Safe Handling of Organic Peroxides: An Overview." *Plant/Operations Progress,* 8(4), 1989.
Nugent, D. P., *Directory of Fire Tests Involving Storage of Flammable and Combustible Liquids in Small Containers.* Society of Fire Protection Engineers, Boston, MA, February, 1994.
Nugent, D. P., "Fire Tests Involving Storage of Flammable and Combustible Liquids in Small Containers." *Journal of Fire Protection Engineering,* 6(1), 1–9, 1994.
Nugent, D. P., "Full Scale Sprinklered Fire Tests Involving Storage of Flammable and Combustible Liquids in Small Containers." Proceedings: International Conference on Fire Research and Engineering, Orlando, FL, September, 1995.
Scheffey, Joseph L. and Tabar, David C., "Hazard Rating System for Flammable and Combustible Liquids." *Process Safety Progress,* 15(4): 230–236, 1996.

3 Administrative Controls

3.1 Synopsis

Safe warehousing of chemicals begins with an explicit management commitment to employees, community and other stakeholders, and includes efforts to protect against the hazards posed by the materials stored and transported to and from the facility. This chapter outlines some of the specific actions management should take to fulfill this responsibility.

3.2 Safety and Risk Management Policies

Operating procedures, safety rules, employee training, emergency preparedness and other management initiatives should be based on a management philosophy which make safety and risk management a primary management responsibility and concern. Policies which derive from this philosophy should address:

- The importance of employee and community safety, environmental protection, and property conservation as core management values,
- The responsibility of each employee to follow all safety practices, and to inform management of unsafe conditions, and,
- The commitment of management to be in compliance with all applicable safety, health, and environmental protection regulations.

It should be noted that the word "safety" has been used in this book to refer to the overall safe operation of the warehouse, as it relates to people, the environment, property and business opportunity.

Employee safety and health policy issues are outlined in Chapter 4.

3.3. Hazard and Risk Management

Identification of hazards which might threaten employee or public safety, the environment or property is the first step. Later chapters in this Guideline discuss the major types of hazards which chemical warehousing might present, and which should be evaluated.

After identifying existing or anticipated hazards, the next step is risk assessment, that is, evaluating and interpreting the hazards. Several widely used hazard identification and risk assessment techniques are described in the literature, including AIChE/CCPS's *Guidelines for Hazard Evaluation Procedures*. Emergency planning is also covered in Chapter 9.

3.4. Control of Ignition Sources

Control of ignition sources is an essential strategy for minimizing the potential for fires and explosions in chemical warehouses. There are several types of ignition sources that could potentially cause a fire in a chemical warehouse; some examples, preventive measures and useful references can be found in Table 3.1. Operations, such as, filling, dispensing, and repackaging of chemicals are sometimes performed in conjunction with chemical warehousing. These operations can also create an ignition potential through static electricity buildup. However, this issue is beyond the scope of this guideline.

3.5. Regulatory Compliance

Companies which handle hazardous chemicals operate in a complex regulatory environment, often subject to many regulations, standards, rules, and procedures from government, standards organizations, and insurers. These requirements can be confusing, redundant or even mutually inconsistent. Moreover, even complete regulatory compliance will not assure a safe warehouse operation. The unique operation of every company therefore requires that management should look beyond the letter of the law to the specific needs of that operation to establish an effective safety policy that is based on the nature of the hazards involved.

Management is best able to identify and respond to risks most effectively when there is a definite commitment to understanding the risks associated with the business, to adequately defining and carrying out a comprehensive safety program, and to funding and otherwise supporting that program.

3.5. Regulatory Compliance

TABLE 3-1
Typical Ignition Sources

Type	Examples	Preventive Measures	References
Open flames/Sparks	Cutting and welding Metal grinding Concrete chipping	Hot work policy	NFPA 51B NFPA Fire Protection Handbook IRInformation
	Smoking	No smoking policy	
	Direct-fired heaters	Isolate or use other type of heater	
Electrical	Arcing and sparking from electrical equipment, including industrial trucks	Install appropriate electrical equipment and appropriate industrial trucks	NFPA 70 NFPA 505
	Lightning	Install lightning protection system	NFPA 780
Hot surfaces	Ducts, pipes, radiant heaters, or industrial trucks exposing heat sensitive material	Provide appropriate space separation and appropriate industrial trucks	MSDS NFPA 505
	Slipping conveyor belt on a roller	Provide inspection and maintenance	Manufacturer's recommendation
Chemical decomposition	Heating of organic peroxide formulations requiring refrigeration	Provide adequate temperature controls	MSDS
Spontaneous ignition	Rags soaked with unsaturated oils such as linseed	Adequate housekeeping practices	MSDS

3.6. Risk Management Organization

Effective risk management involves virtually all functions within a warehousing organization, including personnel, materials handling, maintenance, marketing and finance. Because many issues cross functional lines, a safety and risk management team should be formed with all functions represented. Because of their first-hand knowledge of problems and practical solutions, non-supervisory employees (operators, mechanics, etc.) should be included in the process.

3.7. Employee Hiring, Training and Operations

Most safety policy is carried out at the operations level. Effective screening, training and oversight of employees will result in increased awareness of safety issues and fewer accidents. See also Chapter 4, Employee Safety and Health..

3.7.1. Employee Hiring

Safety related criteria for use in choosing candidates for typical chemical warehouse operations (e.g., including shipping/receiving, put and pick, material handling, etc.) might include:

- a literacy level adequate to be able to read and understand instructions and Material Safety Data Sheets (MSDS)
- lifting strength, as necessary
- warehousing/industrial truck operations experience
- chemical handling experience
- computer use experience, as needed when dealing with inventory tracking systems

Safety and personnel policies of the warehouse should state clearly that the safety of workers in the warehouse depends upon maintaining a drug and alcohol-free workplace.

3.7.2 Training

Hazard communication and personal protective equipment and emergency procedures training are addressed in Chapter 4. As with conventional warehouse operations, employees should be trained in safe lift truck operations, lifting and other procedures which are not specific to chemical warehouse operations. However, in a chemical warehouse, instructions on these routine functions should be tailored to address the unique hazards present in the chemical warehouse environment.

3.7.3. Operations

Operations should be analyzed to identify possible accident scenarios, and measures taken to limit those exposures. Such measures include but are not limited to:

- eliminating or controlling filling, dispensing and sampling operations, particularly of flammable or toxic materials in the warehouse area.
- setting a policy whereby employees work in pairs with high risk materials where exposure can result in a situation immediately dangerous to life and health or that may impact the surrounding community.
- employees working in pairs with high risk materials should be physically separated by enough distance so that a single event will not result in exposure to both employees.
- rotating tasks and employees. Under certain circumstances, long hours of repetition can lead to reduced attention and thus to accidents. Repetitive motion may also lead to cumulative trauma disorders such as carpal tunnel syndrome.
- scheduling regular and routine maintenance of building and equipment, particularly safety equipment. Properly maintained equipment is safer and more efficient to operate.

3.8. Housekeeping

As simple and basic as good housekeeping practices may seem, a commitment to maintaining good order will limit the chance or severity of accidental spills, releases or fires. Good housekeeping practices will also increase employee efficiency and safety.

3.9. Inventory Management

Maintaining tight inventory controls in chemical warehousing is essential to assure compliance with storage arrangement restrictions, such as separation of incompatible materials, storage height, and quantity limits. An up-to-date inventory is also critical information in case of a fire.

Automated inventory identification and management (automated ID) systems have been developed which greatly facilitate inventory control. Automated ID systems have created dramatic changes in the management of chemical warehouses, improving productivity, efficiency, quality control, inventory management, warehouse space utilization, retrieval systems, and safety. The primary reason, from a safety perspective, for an automated ID system in a chemical warehouse is that it makes the task of material handling easier and more efficient, reduces potential for incompatible storage, and avoids exceeding safe storage heights and quantity limitations. While there are clear benefits to safety from the use of Automated ID systems, it should be noted that, like all such systems, they are not infallible. Common sense should not be ignored.

Central to the development of an automated inventory management system is the database of inventory which can contain the following information:

- commodity type and hazard classification;
- assignment of warehouse location information to assist "put and pick" operations;
- MSDS sheets; and
- inventory accounting.

Where bar code labels are generated and applied on site, a unique inventory number can be generated and applied to the specific carton, pallet load, or packaging system. The system should be engineered so that bar code labels correspond with storage standards. The inventory number can then be assigned specific warehouse location information.

Information regarding the volume of inventory at each specific warehouse bin location can be included in the parameters programmed into the database as shipments arrive. Total facility limits on quantity of specific materials or material classes should be considered in accepting materials for storage. Information regarding specific storage requirements related to properties and hazards can be considered by the database in assigning location. Materials to be stored should be evaluated against warehouse design specifications to assure that they are within the facility limitations. For example, the system can be used to direct certain commodities to sections of the warehouse where special fire protection schemes have been installed. Additionally, parameters dealing with compatibility, storage height, and quantity limits can also be addressed. Material to be stored should be evaluated against warehouse specifications to assure that it is within the facility specifications or limitations.

As shipments move though the warehouse, terminals at the various picking stations carry the information which was entered at the receiving dock. Real time interface between existing inventory and the database helps maintain efficiency and quality control.

As items are pulled from the shelves, the bar code is scanned, and the retrieval is downloaded into the database. The inventory accounting and warehouse space allocation are adjusted accordingly. MSDS sheets, customer invoices, and bills of lading can be automatically generated.

In the case of a fire, the inventory management system can inform emergency responders on the amount of each chemical involved. Thus, access to inventory information off-site and in a timely manner should be made available.

3.10. Management of Change

Change is pervasive, a part of our lives. But change in the chemical warehouse creates potential risks which must be evaluated. Change may be gradual and slow, and thus go unnoticed; planned and scheduled; or abrupt and demanding. It may originate outside of the warehouse and affect the entire community where the warehouse is located, or it may be limited to the warehouse or even a single employee, product, or piece of equipment in the warehouse. Management of change involves a process from recognition of change, preferably before it occurs, assessment of the impact of the change on safety, to action to control the impact when appropriate.

Communication at all levels, including management and employees, will cultivate this awareness. Communication with local fire department officials, insurance carriers, cartage companies, suppliers, customers, or the community may also be required .

By maintaining a commitment to loss prevention, management can assess the real risks in any change and develop strategies to deal with them. Loss control professionals can assist management in determining loss potential and identifying solutions.

Documented procedures and parameters for the safe operation of the warehouse are important tools in managing change. Even minor changes in maintenance procedures such as the temperature in the warehouse may have serious consequences.

Staff changes require an awareness of the potential impact on operations. Temporary leave, such as vacation, labor disputes, or illness, may involve employee shortages or untrained temporary laborers. Increased supervision or establishing training requirements for employees temporarily assigned to different duties may be necessary in such cases. Management should assure through documentation of procedures and training that no voids are left in institutional knowledge or assigned roles when employees leave.

Changes in occupancy, such as the types of chemicals and associated hazards, repackaging and transfer activities, product packaging, or method of storage can create new risks.

Such changes in occupancy may necessitate:

- upgrading the warehouse sprinkler system;
- installation of a treatment system or cylinder containment vessel when storing toxic compressed or liquified gases;
- new storage arrangements;
- separation of incompatible chemicals;
- additional staff training;
- fire department notification;
- neighboring occupancy notification;
- changes in the inventory database;
- additional MSDS;
- changes in the emergency plan (see Chapter 9);
- expansion or modification of the facility.

When designing a fire protection system, management should consider not only the company's current operations, but also plans for the facility and future operational needs. If changes in occupancy or storage arrangements are likely, it is usually less costly to anticipate these changes in the original design, rather than retrofit later.

References

Center for Chemical Process Safety. *Guidelines for Hazard Evaluation Procedures.* New York, 1992.

Industrial Risk Insurers, *IRInformation—Guidelines for Loss Prevention and Control.* Hartford, CT

National Fire Protection Association, "Standard for Fire Prevention in Use of Cutting and Welding Processes," NFPA 51B, Quincy, MA, 1994

National Fire Protection Association, "National Electric Code," NFPA 70, Quincy, MA, 1996

National Fire Protection Association, "Fire Safety Standard for Powered Industrial Trucks Including Type Designations, Areas of Use, Conversion, Maintenance, and Operation," NFPA 505, Quincy, MA, 1996

National Fire Protection Association, "Standard for the Installation of Lightning Protection Systems," NFPA 780, Quincy, MA, 1995

National Fire Protection Association, *Fire Protection Handbook,* 18th ed. Quincy, MA, 1997.

Additional Reading

Mansdorf, S. Z., *Complete Manual of Industrial Safety*. Prentice-Hall, Englewood Cliffs, NJ, 1993.

National Safety Council, *Accident Prevention Manual for Industrial Operations Administration and Programs,* 10th ed. Itasca, IL, 1992

4 | Employee Safety and Health

4.1. Synopsis

The safety and health of employees is one of the most important considerations in a chemical warehouse operation. All of the typical nonchemical warehouse related issues such as manual material handling/ergonomics, walking surfaces, and industrial truck operations apply to chemical warehouses. In chemical storage warehouses, because of the inherent hazards of many chemicals, certain elements must be emphasized in the employee safety and health program, including administration and engineering controls, hazard communication, personal protective equipment, and emergency spill response.

4.2. Policy

A written policy should govern safety and health of chemical warehouse employees. The policy should include:
- A statement of management's commitment to employee safety and health.
- The company's definition of safety.
- Goals and scope of the policy.
- Identification of individuals responsible for various elements of the safety program, and their specific responsibilities.

Where use of personal protective equipment (PPE) is required, the policy should explain why PPE is to be used, circumstances where PPE is to be used, and any exceptions and limitations.

4.3. Administrative and Engineering Controls

Effective control of operations—the execution of activities in accordance with established procedures and designs—is essential to safety. Assuring proper execution is a fundamental management responsibility. Most accidents are the result of improper or inadequate execution and thus reflect on management effectiveness. Anticipating the possibility of occasional breakdown in primary controls, procedures and designs frequently provide secondary controls, such as PPE and emergency spill equipment and procedures.

4.3.1. Administrative Controls

Administrative controls include policies and procedures which address the human elements which can impact the frequency of accidents and injuries. Supervision, personnel selection, training, and scheduling policies are examples of administrative controls.

Supervisors should be role models and play an integral part in implementing, gaining and maintaining employee investment in consistent adherence to the safety policy. While upper management is ultimately responsible for safety, the line supervisor is directly responsible for the safety of the employees who report to him or her. He or she is the link between the policies established by management and the employees who must adhere to them on a day-to-day basis. Further, he or she should be a conduit for channeling the ideas of line personnel back to management to further refine and enhance the company's safety program.

Careful screening of applicants during the hiring process is the first step in developing a competent, well-trained work force. New employee physicals, substance abuse screening, and skill testing can help identify individuals with the aptitude, literacy skills, and mental and physical capacity needed to handle a particular job and to understand the hazards associated with that job.

Training is central to management control. The hazards of certain chemicals, as well as regulatory requirements associated with those hazards, pose a more demanding training regimen than for warehouses storing non-chemical commodities. The training program should include properties and hazards of chemicals and precautions to take with each. New employees should always receive training as soon as possible after beginning work and should only work under direct supervision until fully trained. Records of training should be maintained.

4.4. Hazard Communication

Accident reports should be maintained and analyzed as required by regulation, insurance providers and as a means of identifying training or procedure deficiencies.

Supervisors are key persons in the employee safety and health program. Consequently, they should be trained as trainers on the principles of safe operations as they apply to the chemical storage facility. They should monitor their area for human, situational, and environmental factors which affect the safety and health of workers.

4.3.2. Engineering Controls

Engineering controls are best implemented during design and layout of a warehouse and before equipment is specified. As with all warehouse operations, chemical warehouses should address lighting, material handling aids and warehouse layout issues such as aisle width and traffic patterns. Some examples of additional engineering controls to consider for a chemical warehouse are:

- implementing automated product identification systems for inventory control. Bar code systems can code incoming products with information regarding composition, compatibility, storage, location, and quantity. (See Chapter 3.9 for further details on utilizing automated product identification systems for safety.)
- installing detection alarms for materials with poor olfactory warning properties or those that desensitize the olfactory sense, or to alert staff to levels that approach the material's TLV or PEL.
- designing emergency ventilation systems to capture fugitive emissions of toxic, corrosive, or malodorous gases or vapors.

Systems provided primarily for building, product, or environmental protection may also enhance employee safety. Such systems may include sprinkler systems, security systems, fire extinguishers, secondary containment systems, and heating and refrigeration systems.

4.4. Hazard Communication

Hazard communication programs, commonly referred to as HazCom programs, help create a safer working environment and are required by law. These programs disseminate information about the hazards of all chemicals stored in the warehouse to employers and employees. 29 CFR 1910.1200, Hazard Communication Standard, requires employers to develop and implement written HazCom programs.

Chemical manufacturers and importers are responsible for reporting the known hazards of the chemicals they produce, including hazards associated with health, physical properties, and reactivity. This information is provided in the form of Material Safety Data Sheets (MSDSs) and labeling. The information must be made available to employees of the warehouse.

4.4.1. Labels

Labeling is applied by the manufacturer or importer and chemical warehouses are generally not required to conduct labeling operations. While the information found on labels is important to safe warehouse operations, labels may not provide sufficient information to be the sole sources of hazard information.

Chemical warehouse operators should ensure that labels on incoming containers are not removed or defaced. Labels should face the aisles. Operators should make provisions for relabeling containers when necessary. Each label should contain the identity of the chemical, the appropriate hazard warning, and the address and phone number of the manufacturer.

The American National Standards Institute (ANSI) has established a voluntary standard for chemical labeling, ANSI Z129.1 1994, "Hazardous Industrial Chemical Precautionary Labeling," which includes readily identifiable symbols for poisons, corrosives, flammables, and explosives. Irritants, combustible liquids, pyrophoric chemicals, oxidizers, sensitizers, physiologically inert vapors, gases and other hazardous materials are also addressed.

4.4.2. Material Safety Data Sheets

The Material Safety Data Sheet (MSDS) should arrive at the chemical warehouse from the manufacturer or importer with the first shipment and whenever the MSDS is revised. As with labels, chemical warehouse employees are not required to know how to write a MSDS, but must be trained in how to understand their contents.

All employers, including chemical warehouse operators, must keep an accurate and up-to-date MSDS for each hazardous chemical in the work place. Although the formats may vary, MSDSs usually contain the following information:

- Identity (as used on container label and hazardous chemical inventory form).

4.4. Hazard Communication

- Specific chemical identity and common names.
- Physical and chemical characteristics.
- Physical hazards and extinguishment procedures.
- Other substances that may react with the chemical.
- Environmental conditions that may result in a fire or explosion.
- Known acute and chronic health effects and related health information.
- Primary route(s) of entry.
- Exposure limits.
- Whether the chemical is considered to be a carcinogen by the National Toxicology Program (NTP), International Agency for Research on Cancer (IARC), or OSHA.
- Spill and leak disposal procedures.
- Precautionary measures.
- Emergency and first aid procedures.
- Date of preparation or last change.
- Identity of organization responsible for preparing the Material Safety Data Sheet (MSDS).

Under HazCom, a chemical warehouse operator is responsible for making the MSDS available to employees in a readily accessible format and for providing training to inform employees on how to read and understand a MSDS.

4.4.3. Employee Information and Training

A hazardous chemical information and training program is an important element of hazard communication. It should begin with new employees on their first day. If new hazards are introduced into the warehouse, additional information and training should be provided to employees. Employees should be told the location and availability of the MSDS's. Employee training should include a discussion on the hazardous chemicals in the warehouse. This training should also include the physical and health hazards of the hazardous chemicals that are stored, as well as measures that can be taken for protection. Finally, training should explain the appropriate use of the container labeling system and MSDSs.

It is the supervisor's job to assure that employees understand the hazards associated with the materials stored and handled. Supervisor safety training can be performed in-house or by professional organizations such as the National Safety Council (NSC), the Insurance Institute of America (IIA) or independent training organizations.

4. Employee Safety and Health

Employees also require refresher training from time to time. Monitoring of employee activities and periodic questioning of employees should be utilized to determine the ongoing level of employee awareness.

4.5. Personal Protective Equipment

Routine operations at chemical warehouse facilities do not present direct exposure to health hazards of chemicals handled in sealed containers. Extra precautions are required for accidental releases. The proper selection and use of PPE can protect employees from the effects of chemicals in the event of a release. The effects of chemical exposure can range from acute trauma, such as skin irritations and burns, to chronic degenerative diseases.

Appropriate PPE, such as safety helmets and gloves, can also be effective against mechanical hazards such as falling objects and sharp instruments.

4.5.1. Implementing a PPE Program

The initial step in the selection process is an analysis of the potential hazards present in each job, followed by selection of appropriate PPE.
- Develop a written policy and effectively communicate it to employees. The components of a PPE program should be contained in a company's safety policy as outlined in Section 4.2.
- Select the proper types of PPE based on a hazards assessment and the suitability of PPE to protect from each potential hazard.
- Implement an effective training program for the use and maintenance of PPE.
- Strictly enforce the PPE policy.

4.5.2. Selecting PPE

The initial step in the selection process is an analysis of the potential hazards present in each job, followed by selection of appropriate PPE.

Some of the criteria for selection of chemical protective clothing, footwear, head, eye, face, hand and respiratory protection follow:

4.5. Personal Protective Equipment

4.5.3. Chemical Protective Clothing

Chemical warehouse workers normally wear chemical protective clothing only in spill response situations. Chemical protective clothing is used to protect the skin and eyes against gaseous, liquid and particulate chemical hazards. Although certain types of chemical protective clothing may be gas or vapor resistant, it is not intended to provide respiratory protection. Chemical protective clothing includes an ensemble composed of handwear, footwear, aprons and full body protection. Materials commonly used to make chemical protective clothing include various natural or synthetic rubber and plastic compounds and other fabrics coated with these materials. Certain chemical protective clothing may be designed for repeated uses while others may be designed to be disposed of after a single use or exposure.

Manufacturers or importers should provide information regarding the chemical resistance for particular products or material components. Table 4-1, "Chemical Protection Clothing—Chemical Resistance Chart," lists the chemical resistance rating for some representative chemicals previously referenced in Chapter 2.

Chemicals will pass through any protective barrier given sufficient time. Tests performed in accordance with the American Society for Testing and Materials, ASTM F739, 96 and ASTM F903, 96 measure a material's resistance to a chemical by determining its breakthrough time and permeation rate. A third criteria, degradation rate, measures the changes to the chemical barrier when it comes into contact with the material. This may include swelling of the material. As a long term performance characteristic, degradation can also occur due to tears or abrading the material or other mechanical damage. Suits intended for reuse should be tested and requalified for service in accordance with NFPA 1991, "Standard on Vapor-Protective Suits for Hazardous Chemical Emergencies," and NFPA 1992, "Standard on Liquid Splash-Protective Suits for Hazardous Chemical Emergencies."

Chemical protective clothing should be chosen with consideration given to the tasks that will be performed, the expected duration of contact, the breakthrough time, permeation rates and degradation rates for the particular chemicals being handled.

When selecting protective clothing consider that:

- Permeation of a chemical is not always visible.
- Materials that protect against one chemical may not protect against another.
- No single material protects against all chemicals.

TABLE 4-1
Chemical Protective Clothing—Chemical Resistance Chart

Chemical	LDPE	HDPE	PP/PA	PMP	PC	PVC	PSF	TFE/FEP
Ammonia	EE	EE	EE	EE	NN	EG	GF	EE
Chlorine	GN	EF	FN	GN	EG	EE	NN	EE
Propane	NN	FN	NN	NN	FN	EG	FF	EE
Acetone	GG	EE	EE	EE	NN	NN	NN	EE
Acrylonitrile	EE	EE	FN	FN	NN	NN	NN	EE
Mineral Spirits	GN	EE	EE	EG	EG	EG	EE	EE
Calcium Hypochlorite	EE	EE	EE	EG	FN	GF	EE	EE

Materials Key		Ratings Key	
LDPE	Low Density Polyethylene	E	Excellent
HDPE	High Density Polyethylene	G	Good
PP/PA	Polypropylene/Polyallomer	F	Fair
PMP	Polymethylpentene	N	Not Recommended
PC	Polycarbonate		
PVC	Polyvinylchloride	First letter applies to conditions at 68°F (20°C) Second letter to those at 122°F (50°C)	
PSF	Polysulfone		
TFE/FEP	Tetrafluoroethylene/Fluorinated Ethylene-Propylene (Teflon)		

Source: Manufacturers' Test Data. Please note this guide is intended as general information. Since each pair of ratings is for ideal conditions, consider all factors when evaluating chemical resistance.

- Protective clothing may look the same but have different capabilities. Do not depend on color or appearance in choosing clothing.
- Once permeation begins, it will continue until the chemical breaks through the fabric or until such time the fabric is adequately decontaminated.

If clothing is suspected of being contaminated it must either be decontaminated following the material manufacturer's procedures or disposed of in accordance with the U.S. EPA requirements for hazardous waste disposal.

4.5. Personal Protective Equipment

4.5.4. Foot Protection

As in warehouses which store non-chemical commodities, chemical warehouse workers should wear safety shoes and/or boots which provide impact and compression protection. For spill clean up, various rubber and plastic materials are appropriate for the construction of boots to be used in chemical environments. Additional information can be found in American National Standards Institute (ANSI) Z41 1991, "Personal Protection—Protective Footwear."

4.5.5. Head Protection

Use of hard hats should correspond with usage in non-chemical warehouses. Where hard hats are appropriate, ANSI Z89.1 1986, "Personal Protection—Protective Headwear," defines hard hats in three categories. Class A hard hats are designed to protect the head from falling objects and electrical shock during contact with exposed low voltage conductors. Class B hard hats are designed for use where exposure to high voltage exists. Class C hard hats are designed for protection against impact from falling objects

4.5.6. Eye and Face Protection

Considerable variety is available in the types, styles, and applications of eye and face protection. For the conditions which exist in most chemical storage facilities, where packaging remains sealed, eye protection will be dictated by the presence of physical hazards. Many warehouse operators require safety spectacles in warehouse and maintenance areas. Persons who wear prescription lenses should wear protective eyewear fitted with prescription lenses or protective devices worn over prescription eyewear.

4.5.7. Hand Protection

During routine operation in a warehouse, gloves that protect hands during manual handling should be chosen. For spill clean up, chemical protective gloves must be available. Gloves should be selected for their performance characteristics relative to specific hazards. Such hazards include cuts, abrasions, burns, and skin contact with chemicals. Materials commonly used in the manufacture of gloves include neoprene, butyl, polyvinylchloride (PVC) and nitrile. Manufacturers can provide documentation stating the glove's compatibility to certain chemicals or chemical classes, as well as permeation rates and breakthrough times for the glove material.

4.5.8. Respirators

Chemical releases can contaminate the work atmosphere and may pose respiratory hazards. Respiratory hazards include particulates (dusts), vapors, gases, and fumes. Respirators are never considered the primary protection for employees. Under normal operations in a chemical warehouse where no packaging or processing occurs, respirators need not be part of routine PPE use. However, it may be necessary to use respiratory protection when performing clean-up of chemical releases or escape from a location within a warehouse atmosphere that has chemical exposure levels above TLV or PEL, or that are approaching IDLH levels. Like other PPE, respirators should be chosen for their ability to protect the employee from the hazards of a release of the particular chemicals handled in that facility. In general, respirators for chemical warehouses should be chosen to reflect short term exposures for escape or clean up of a spill, rather than a long-term exposure during a routine work day. If respirators are to be used in a chemical warehouse, the warehouse operator must have a site-specific written respirator program which includes training of all users.

29 CFR 1910.134, "Respiratory Protection," contains requirements for personnel who are required to wear respiratory protection. The requirement states that when engineering controls are not feasible, or while they are being instituted, appropriate respirators shall be used pursuant to the following requirements:

1. Respirators shall be provided by the employer when such equipment is necessary to protect the health of the employee.
2. The employer shall provide respirators what are applicable and suitable for the purpose intended.
3. The employer shall be responsible for the establishment and maintenance of a respiratory protection program.

The respiratory protection program prescribed by OSHA contains provisions for the following:

- Written standard operating procedures.
- Respirators selected on the basis of hazards.
- Instruction and training of the user.
- Cleaning and disinfection.
- Storage and inspection.

4.5. Personal Protective Equipment

- Surveillance of work area conditions.
- Evaluation of the respiratory protection program.
- Medical review and evaluation of employees who wear respiratory protection.
- Use of Mine Safety and Health Administration (MSHA) and/or National Institute for Occupational Safety and Health (NIOSH) approved respirators.

Types of respirators include:

- *Self Contained Breathing Apparatus (SCBA)*: Where toxic contaminants or levels are unknown, the material has poor warning properties, or the atmosphere is at or has exceeded IDLH limits, SCBA is required. An IDLH atmosphere is one where conditions pose an immediate threat to life or health or conditions that pose an immediate threat to severe exposure or contamination. If an atmosphere exceeds IDLH level, the only respiratory protection method allowed is SCBA. SCBAs are one form of supplied air respirators.
- *Air-line Respirators:* These are respirators similar to SCBA except that the unit is connected to a hose that can be deployed to a maximum distance of 300 feet. These respirators can only be used in IDLH atmospheres when they are equipped with an escape air supply that can provide breathing air for at least 5 minutes. This limitation is necessary because most air-line respirators are entirely dependent upon an air supply that is not carried by the wearer of the respirator.
- *Air Purifying Respirators:* These devices cleanse the contaminated atmosphere using chemical or mechanical filters individually or in combination. Chemical filters afford protection against concentrations of 10–1,000 PPM by volume, depending on the contaminant. These cannot be used in IDLH atmospheres or atmospheres that are oxygen deficient. Furthermore, they cannot be used for chemicals that have poor warning properties.

The effectiveness of the air purifying respirator depends on choosing the proper filter, cartridge or canister for the type of exposure present. Air purifying respirators rely on the use of activated charcoal or other chemical-specific filters, which remove gases and vapors, or filters which remove particulates such as dusts. Others utilize a combination of both.

Cartridges are designed for limited use and should be periodically replaced in accordance with manufacturers recommendations or more restrictive company guidelines. Some cartridges utilize an external color coded indicator which alerts the user that breakthrough is near and the cartridge should be replaced.

Replacement cartridges and filters come in a wide variety and should be matched to protect the worker from a particular contaminant. Reference information for selection and identification of air purifying canisters and cartridges can be found in 29 CFR 1910.134, "Respiratory Protection." This color coding and labeling system enables users to select cartridges which are best suited for a particular contaminant. The basic color coding system is shown in Table 4-2.

There are additional requirements when using these filters in combination for protection against multiple contaminants, and most distributors of this type of equipment have technical representatives manning toll free telephone numbers to assist users with proper selection.

4.5.9. Respirator Selection

The maximum potential concentration of chemical contaminant should be used as the basis for respirator selection in the chemical warehouse. For accidental releases, where concentration cannot be predicted, a SCBA must be used.

4.5.10. Respirator Usage

Employees must be trained in the proper use of respirators. Only persons who are physically capable of wearing respirators and of performing the clean up functions for which the respirator must be worn should be assigned tasks which carry the potential to require respirator usage. A medical evaluation should be performed to verify the employee's physical condition. Respirators should be fitted to each individual and potential effects of facial hair should be considered. Respirators should be cleaned and sanitized regularly after each use. Respirators should also be inspected and maintained according to manufacturer's instructions. Worn and defective parts should be replaced immediately.

4.5.11. Training

When workers are required to wear PPE, an effective training program is necessary to ensure proper use. Employees must demonstrate an understanding of the hazards involved and the ability to use PPE in the manner it was intended. A PPE training program should:

- Describe what hazardous conditions are present in the work place.

4.5. Personal Protective Equipment

TABLE 4-2
Color Code for Cartridges/Canisters

Color Assigned	Atmosphere Contaminated
White	Acid Gases Only
White with ½-inch Green Stripe completely around the canister near the bottom	Hydrocyanic Acid Gas
White with ½-inch Yellow Stripe completely around the canister near the bottom	Chlorine Gas
Black	Organic Vapors Only
Green	Ammonia
Green with ½-inch White Stripe completely around the canister near the bottom	Acid Gases and Ammonia
Blue	Carbon Monoxide
Yellow	Acid Gases and Organic Vapors
Yellow with ½-inch Blue Stripe completely around the canister near the bottom.	Hydrocyanic Acid Gas and Chloropicrin Vapor
Brown	Acid Gases, Organic Vapors, and Ammonia
Purple (Magenta)	Radioactive Materials, except Tritium and Nobel (Inert) Gases
Canister Color for Contaminant, as designated above, with ½-inch Gray Stripe completely around the canister near the top.	Particulates (Dusts, Fumes, Mists, Fogs, or Smoke) in combination with any of the above gases or vapors
Red with ½-inch Gray Stripe completely around the canister near the top.	All of the above atmospheric contaminants
Source: 29 CFR 1910.134, "Respiratory Protection"	

- Explain what can or cannot be done about hazardous conditions in the work place.
- Explain why a certain type of PPE has been selected.
- Explain the capabilities and limitations of the PPE.
- Demonstrate how to fit, adjust and use PPE properly, and practice in using it.
- Explain the written company policy and enforcement procedures.
- Explain how to deal with emergency situations.

- Explain maintenance, cleaning, repairing and replacing PPE.

Once employees have been trained in the proper use of PPE, periodic audits will be needed to confirm that the equipment is consistently and properly used. Furthermore, PPE should remain in the work place. It should never be taken home.

Annual or periodic refresher training is recommended to address:

- Changes in the work place or processes.
- Changes in the types of PPE used.
- Inadequacies in an employee's use and knowledge of assigned PPE.

4.5.12. Maintenance and Inspection

In order to be effective in controlling exposure, PPE needs to be maintained. Proper maintenance procedures will include:

- Periodic inspection and maintenance of all PPE, according to manufacturer's recommended inspection and maintenance procedures.
- Recordkeeping regarding all selection, inspection and maintenance activities.
- Equipment repair with the use of manufacturer's parts for repair, and using qualified repairmen.
- Cleaning, disinfecting and/or decontaminating PPE as needed. PPE which cannot be decontaminated should be disposed of properly and in a manner that protects employees from exposure to hazards.

4.6. Safety Equipment

Additional equipment which may be required for use in the event of a chemical release includes:

- two-way communication systems
- panic hardware/deadman switches
- eye wash stations/wash basins/deluge showers
- evacuation alarms
- first aid kits
- spill carts
- building exit route maps
- emergency lighting
- air sampling equipment

Eye wash stations should be strategically located.

Hand and face wash basins can minimize the transfer of chemicals to other body parts and thus reduce the threat of poisoning.

Deluge showers should be located near points of egress from areas of potential exposure. They should provide a minimum flow of 30 gpm.

Air sampling equipment may be useful in determining the level of toxicity after a spill. Where air sampling equipment reveals an IDLH environment, entry should be restricted to SCBA respirators.

4.7. Emergency Response Training

Emergency response to spills, fires, or medical emergencies may expose employees to safety and health hazards. To assure employee safety as well as effective action, employees assigned emergency response responsibilities should receive appropriate periodic training.

4.7.1. Emergency Spill Response

To provide a mechanism for rapid response to spills or releases of hazardous materials, facility management may elect to provide an emergency spill response team. Such a group is delegated the responsibility of mitigating accidental chemical releases. This may be accomplished by outside contractors who provide the service or by employees specifically trained for these tasks. Regardless of the strategy selected, facility management needs to be aware of OSHA and EPA regulations which address the requirements for persons who perform emergency response operations for the release of hazardous materials.

The OSHA and EPA regulations contain the following specific requirements for emergency response operations.

1. Development of a safety and health plan designed to identify, evaluate and control health and safety hazards. This includes providing for emergency response.
2. An evaluation of the site's characteristics prior to entry by trained persons to identify potential site hazards and aid in the selection of appropriate employee protection methods. Included would be all suspected conditions that are IDLH or that may cause serious harm.
3. Implementation of a site control program to protect employees against hazardous materials contamination. At a minimum, the site safety plan must designate the location of the spill or release site, site work zones, site communications, safe work practices

and identification of the nearest medical assistance. This includes the use of the "buddy system" that requires at least two persons always enter into a contaminated area and that a backup team of at least two persons is provided for rescue purposes.
4. Training of employees before they are allowed to engage in emergency response that could expose them to safety and health hazards. Specific training levels are listed for clean-up personnel, equipment operators, supervisory employees and for the various levels of emergency response personnel.

Different levels of training are required depending on the duties and functions of each responder plus demonstrated competence or annual refresher training sufficient to maintain competence.

✓ *Awareness Level:* These are individuals likely to witness or discover a hazardous materials release and initiate the emergency response. Persons trained to this level must demonstrate competency in such areas as recognizing the presence of a hazardous material in an emergency, the risks involved, and the role they should play.

✓ *Operations Level:* These are individuals who respond for the purpose of protecting property, persons, or the nearby environment without actually stopping the release. Persons trained to this level must have eight hours of training plus "Awareness Level" competency.

✓ *Hazardous Materials Technicians:* These individuals respond to stop the release and they must have 24 hours of training equal to the "Operations Level" and must demonstrate competence in several specific areas.

✓ *Hazardous Materials Specialists:* These individuals support the technicians but require a more specific knowledge of the substance to be contained. They require 24 hours of training equal to the technician level and must demonstrate competence in certain areas.

✓ *Incident Commander:* The incident commander assumes control of the scene beyond the "Awareness Level". They must have 24 hours of training equal to the "Operations Level" and must demonstrate competence in several specific areas.

5. Medical surveillance at least annually and at the end of employment for all employees exposed to any particular hazardous substance at or above established exposure levels or those who wear respiratory protection for 30 days or more per year. Such surveillance is also required if the worker is exposed by unex-

4.7. Emergency Response Training

pected releases of chemicals under emergency response conditions.
6. Air monitoring to identify and quantify levels of hazardous substances with periodic monitoring to assure that proper protective equipment is being used.
7. Implementation of a decontamination procedure before any employee or equipment may enter into or leave an area of potential hazardous exposure; operating procedures to minimize exposure through contact with exposed equipment, other employees, or clothing; and provide showers and change rooms where needed.
8. Implementation of a site safety plan prior to beginning emergency response operations. Such plans must address:
   ✓ Personnel roles and assignments;
   ✓ Lines of authority, training and communications;
   ✓ Site security;
   ✓ Safe places of refuge;
   ✓ Evacuation routes and procedures;
   ✓ Emergency medical treatment; and
   ✓ Alerting of employees of an emergency.

A site safety plan should graphically indicate the areas where the actual emergency response will occur, the location and path to and from the work area, and the location where administrative support functions can be performed. Figure 4-1 illustrates an example of hot, warm and cold zones. The hot zone is the area where personnel will perform the work. Operations in this area will require respiratory protection and some level of chemical protective clothing. The warm zone serves as the area where entry into and egress from the hot zone occurs. This area contains the contamination reduction corridor. Decontamination of chemical protective clothing, equipment and any contaminated persons occurs in this corridor. Depending on the hazardous material, persons working in this area may require respiratory protection and chemical protective clothing. The warm zone also is the location where the backup team will stand by and may contain the area of safe refuge. The cold zone is the area where personnel who are not trained for emergency response can be relocated.

4.7.2. Manual Fire Fighting

If a warehouse owner or operator decides to have an emergency response team, the team should be trained, equipped, and drilled in accordance with 29 CFR 1910.156, "Fire Brigades."

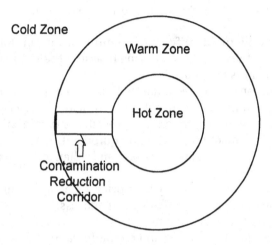

FIGURE 4-1. *Emergency Spill Response Zones*

4.7.3. First Aid

Given the potential for medical emergencies, it would be prudent to have an employee trained to handle such situations. Instruction in first aid and cardiopulmonary resuscitation (CPR) can be obtained from many organizations offering programs sanctioned by the American Red Cross and the American Heart Association respectively.

References

American National Standards Institute, ANSI Z41, "Personal Protection—Protective Footwear," 1991.
American National Standards Institute, ANSI Z89.1, "Personal Protection—Protective Headwear," 1986.
American National Standards Institute, ANSI Z129.1, "Hazardous Industrial Chemical Precautionary Labeling," 1994.
American National Standards Institute, ANSI Z88.2, "Standard Practice for Respiratory Protection," 1992.
American Society for Testing and Materials, ASTM F739, "Standard Test Method for Resistance of Protective Clothing Material to Permeation by Liquids and Gases Under Conditions of Continuos Contact," 1996.
American Society for Testing and Materials, ASTM F903, "Standard Test Method for Resistance of Protective Clothing to Permeation by Liquids," 1996.

National Fire Protection Association, "Standard on Vapor Protective Suits for Hazardous Chemical Emergencies," NFPA 1991, Quincy, MA, 1994.
National Fire Protection Association, "Standard on Liquid Splash-Protective Suits for Hazardous Chemical Emergencies," NFPA 1992, Quincy, MA, 1994.
29 Code of Federal Regulations, Part 1910, Subpart I, "Personal Protective Equipment."
29 Code of Federal Regulations, Part 1910.134, "Respiratory Protection."
29 Code of Federal Regulations, Part 1910.1200, "Hazard Communication Standard."
29 Code of Federal Regulations, Part 1910.156, "Fire Brigades."

Additional Reading

American Society for Testing and Materials, ASTM F1001, "Guide for Selecting Chemicals to Evaluate Protective Clothing Materials," 1989.
American Society for Testing and Materials, ASTM F1052, "Pressure Testing of Gas-Tight Totally Encapsulating Chemical Protective Suits," 1987.
Kamnel, David W., Noyes, Roland T., Riskowski, Gerald L., Hofman, Vernon L.
Midwest Plan Service, "Designing Facilities for Pesticide and Fertilizer Containment," 1991
Mansdorf, S.Z., *Complete Manual of Industrial Safety*. Prentice-Hall, Inc. Englewood Cliffs, NJ, 1993.
National Safety Council, "Accident Prevention Manual for Industrial Operations Administration and Programs," Ninth Edition, Itasca, IL, 1988.
National Fire Protection Association, "Standard on Fire Department Occupational Safety and Health Program," NFPA 1500, Quincy, MA, 1992.

5 Site Considerations

5.1. Synopsis

Local population density and sensitivity, land use and environmental considerations should play an important role in the decision to build, use or continue operation of a warehouse for chemical storage. Local conditions such as high population density, nearby schools or hospitals, or highly sensitive plant and animal species can magnify the problems associated with a chemical release. Natural perils such as earthquakes, floods and hurricanes should be considered as they may precipitate chemical releases. Exposure from nearby industries, such as warehouses, flammable liquid storage or processing facilities should be evaluated. Additionally, the potential for riot and civil commotion, the adequacy of emergency responders, and the adequacy and reliability of utilities such as electricity and water supply should also be addressed.

5.2 Health and Environmental Exposure

Exposure to the public and the environment from certain chemicals may result in serious short and long term consequences. Events that can precipitate a chemical release from a warehouse include fire, on-site accidents, natural perils, or accidents involving trucks making deliveries to or from the warehouse. These events may result in dispersion of hazardous chemicals throughout the surrounding community, particularly if a release involves smoke, vapors, or gases. A hazardous release may involve the actual chemicals stored, reaction products from chemical decomposition, or mixing of incompatible chemicals, or products of combustion.

Dispersion of chemicals into the environment may occur via ground and surface water or sanitary and storm sewer systems. Gases, vapors and combustion products can be dispersed in the air. In addition, particulate matter and condensate can precipitate out of smoke plumes. The risks to the public will be the result of several variables, including quantities, physical properties and hazards, as well as the adequacy and effectiveness of mitigation measures.

5.2.1. Baseline Environmental Assessment

An environmental assessment of the site and surroundings areas and waterways may be needed before acquisition to identify preexisting soil, ground water and waterway contamination. The data developed could help establish liability for prior contamination and, later, to verify the effectiveness of containment and control measures.

5.2.2. Population Proximity, Density, and Sensitivity

The potential consequence of a chemical release from a warehouse located in an area having a high population density is greater than a warehouse located in a sparsely populated area. Additionally, certain segments of the population are more susceptible to chemical exposure related health problems than others. Children, the elderly, and people with preexisting health problems will be more sensitive to low level chemical exposure than young healthy adults.

Therefore, where making a site assessment, locations in heavily populated areas or those that are adjacent to schools, nursing homes and hospitals should be considered high risk areas that may be undesirable. Existing facilities should also be reevaluated periodically since communities that were once considered acceptable may become undesirable due to local changes. Industrial zoned areas are preferable because residential development is less likely to encroach on the site. An area that is considered high risk may become more acceptable by implementing additional preventive and mitigative features into the design and operation of the facility.

5.2.3. Warehouse Truck Traffic

Chemical transportation between major highways and warehouses extends exposure beyond the site and into the surrounding community. Site selection considerations should include traffic patterns relative to population distribution, schools, waterways, railroad crossings, and

5.3. Natural Peril Exposures

similar potentially vulnerable exposures consistent with the chemical hazards that are presented.

5.2.4. Highly Sensitive Environments

Certain areas of the environment are extremely sensitive to certain chemicals, or may present enormous cleanup problems. Wetlands, and rivers and lakes that serve as water supplies, are areas of the environment that are highly sensitive to chemical contamination. The problems created by these environmental features may be exacerbated by topography and soil conditions which speeds the spread of chemical contamination.

The character of the potential contaminants is a key consideration in the environmental vulnerability to a release. Solubility, biodegradability, toxicity and physical state must be considered in evaluating environmental risk.

5.2.5 Surface Water, Ground Water, and Soil Permeability

A chemical release into surface or ground water can present a greater pollution problem than a release onto the surrounding soil. A release into a body of water will become more dispersed in water, particularly if the chemicals are water soluble. This contamination can greatly affect both drinking water supplies and plant and animal life. Unlike a large body of water, contaminated soil can be treated as a hazardous waste. Since waterways are sometimes used as a vital link in the supply/distribution chain, areas adjacent to waterways cannot always be avoided as chemical warehousing sites. They should, however, be considered high risk locations which may require more extensive preventive and mitigative measures.

As part of a site investigation, the proximity of ground or sub-surface water should be identified. Locating a chemical warehouse directly above an aquifer is particularly undesirable and should be avoided, especially if it is a source of drinking water. The permeability of the soil at the site should also be determined.. Highly porous soils, such as sand, are less desirable than denser clay soils. Porous soils will allow chemicals to percolate to greater depths.

5.3. Natural Peril Exposures

Siting considerations pertaining to natural perils, such as earthquake, flood, hurricane, tornado, lightning, and arctic freeze, should include an understanding of the potential exposure. The probability and severity of the exposure will determine the risk to the site. Natural perils such as flood, hurricane, tornado, lightning, and arctic freeze are seasonal. Earthquakes, on the other hand, occur without warning. Some natural perils can precipitate the occurrence of other perils; earthquakes can precipitate landslides and tidal waves or tsunamis.

Regardless of the type of occupancy within a given facility, the risk of exposure from a potential natural peril is constant. However, the impact of a natural peril exposure on a chemical warehouse may be quite different from non-chemical occupancies. If areas subject to the effects of natural perils cannot be avoided, mitigative features presented in this book can be incorporated into the design and operation of the warehouse.

5.3.1. Earthquake

Earthquakes can cause uplift, subsidence and offsetting of the earth. Additionally, liquefaction of soils, landslides, and flooding of coastal areas (tsunamis) and areas adjacent to confined bodies of water (seiche) may also result. The uplifting, subsidence and offsetting of the earth may cause building collapse and damage to fire water and natural gas piping, communication/alarm wiring, and electrical power lines. Collapse of storage arrays due to column buckling and failed connectors can result in chemical spills.

The location of earthquake prone areas can be determined by consulting a seismic zone map. These maps appear in all three of the model building codes used in the United States, namely the BOCA National Building Code, the Standard Building Code, and the Uniform Building Code (UBC), and assign a seismic zone to geographic areas. These maps will not provide a prediction of the earthquake risk but will generally identify earthquake prone areas. It should also be noted that the seismic zone maps appearing in the various model building codes do not assign the same seismic zone to a given area.

Magnitude is a quantitative measurement of the total energy released during an earthquake. In the United States., the magnitude of an earthquake can be measured using a logarithmic scale known as the "Richter Scale."

Earthquake intensity is a qualitative or subjective measurement of the observed effects of an earthquake at a specific location. The scale used in the United States is known as the "Modified Mercalli (MM) Intensity Scale."

5.3. Natural Peril Exposures

5.3.2. Flood

Flood waters may cause damage or release chemicals to the environment. Water damage can result to both the warehouse building and the commodity stored. Cartoned commodity stacks sitting in water may be susceptible to collapse, due to weakening of the cartons. Capillary action may allow water to migrate up into cartoned commodity stored above the water line. In the event a chemical release occurs, pollution to the environment may also result. If the released chemicals are water reactive, they may result in the release of toxic gases or a fire.

The most up-to-date resources for the identification of flood prone areas are flood insurance studies (FIS) and flood insurance rate maps (FIRM). An FIS is used to prepare a FIRM. These resources are published by the Federal Emergency Management Agency (FEMA), a division of the Department of Housing and Urban Development.

A FIRM provides the user a number of important details relative to the potential for flooding in a given area, due to weather related events. Special Hazard Flood Areas (SHFA), the flood insurance zones, 100-year flood elevations (base flood elevations), 500-year flood elevations, and areas designated as a regulatory floodway are indicated on a FIRM.

For an existing or proposed facility, the warehouse, roadways, and the surrounding land can be evaluated for flood potential using a FIRM. The 100-year flood interval and its associated flood elevation are the most widely used criteria for estimating flood potential. The 100-year flood zones are designated as Zones A (non-coastal) or V (coastal) on a FIRM.

5.3.3. Hurricanes

Hurricanes predominantly effect the Gulf and Atlantic coasts of North America. These weather systems can also occur in other regions of the world where they may be identified by other names such as typhoons or tropical cyclones. Hurricanes that effect North America are spawned over the western Atlantic Ocean and usually develop between July and September. These weather systems are very powerful and can affect the weather pattern for hundreds of miles (kilometers) from the center of the storm. The National Weather Service can usually provide a 2–3 day warning to those areas in the expected path of a hurricane.

The effects of a hurricane, when it makes landfall, will be different for coastal and slightly inland areas versus inland areas. Coastal and adjacent areas can be subject to storm surge, high winds, intense rainfall and possible tornadoes. A storm surge can be the most devastating phenomenon associated with hurricanes, and coastal areas are most at risk. Due to the lower atmospheric pressure associated with a hurricane, wind driven waves and shallow ocean depth a storm surge can raise the coastal sea level upwards of 20 ft. (6.1 m) above normal levels. A storm surge will diminish in depth approximately 1–2 ft. (0.3–0.6 m) for every mile (1.6 km.) it travels inland. A period of high tides occurring during a storm surge will intensify the effect.

The intensity of a hurricane will diminish after it makes landfall. The effects of a hurricane on inland areas can include flash flooding and riverine flooding from intense rainfall. Due to the effects of a storm surge on coastal areas, rivers draining into the ocean are susceptible to increased levels causing inland areas to flood.

High velocity winds associated with hurricanes can cause roof damage, the most prevalent type of building damage during one of these events. This damage can include removal of roof components such as perimeter flashing, roof covering and insulation, or the entire roof assembly. High velocity winds over a flat roof will cause a reduced pressure over the roof. This pressure will be lower than the atmospheric pressure under the roof resulting in an uplifting force. The largest differential pressures will occur at the roof perimeter and at the wall corners. The pressure differential will be greater if there is an opening in the windward side of the warehouse. Wind damage to both windows and doors can create these openings. Excessive force can also result in exterior building wall collapse. Wall collapse is dependent on their design and the wind forces to which they are subjected. Heavy rainfall occurring during or after wind damage to a warehouse can result in water damage to the stored commodity and the building.

The Saffir-Simpson Damage-Potential Scale categorizes hurricanes by their intensity. This intensity is measured in terms of their maximum sustained winds and central barometric pressure. Table 5-1 lists the intensity parameters that determine the hurricane categories on the Saffir-Simpson Scale.

5.3.4. Tornadoes

Tornadoes are considered nature's most violent storm and are associated with the occurrence of severe thunderstorms and hurricanes. As opposed to a hurricane, where damage is possible over a wide geographic area, a tornado causes localized damage due to direct contact between objects and the swirling vortex. When a tornado comes in direct contact with a building, the usual result is total destruction due to wind damage.

5.3. Natural Peril Exposures

TABLE 5-1
Saffir-Simpson Hurricane Damage Potential Scale

Level	Maximum Sustained Winds mph (kph)	Central Barometric Pressure in. Hg. (millibars)
1	74–95 (120–153)	>28.94 (980)
2	96–109 (154–177)	28.50–28.94 (965–979)
3	110–130 (178–209)	27.91–28.49 (945–964)
4	131–155 (210–249)	27.17–27.90 (920–944)
5	>155+ (249+)	<27.17 (920)

Tornadoes can be ranked on the Fujita Tornado Wind Intensity Scale. This scale categorizes tornadoes by their vortex wind speed and a property destruction description. Table 5-2 lists the parameters that determine tornado intensity on the Fujita Scale. Although tornadoes have been reported with wind velocities higher than what is shown in Table 5-2, they are rare. The most common tornadoes are categorized as F-1, with the least common categorized as F-5. A "Severe" tornado can have a vortex diameter of 200 yards (183 m), a horizontal travel speed of 30 mph (kph) and a travel path of 2 miles (3.2 km) after making contact with the ground. While warehouses are not usually designed to withstand the forces of a tornado, the warehouse contingency plan should identify areas most suitable for employee safety.

TABLE 5-2
Fujita Tornado Wind Intensity Scale

F- Scale	Wind Speed mph (kph)	Damage
F-0	40–72 (64-116)	Light
F-1	73–112 (118-180)	Moderate
F-2	113–157 (182-253)	Considerable
F-3	158–206 (254-332)	Severe
F-4	207–260 (333-419)	Devastating
F-5	261–318 (420-512)	Incredible

Tornadoes have been known to occur in most of the contiguous 48 states. However, tornadoes occur most frequently in the Gulf Coast and Midwest states with the highest number of tornadoes occurring annually in Oklahoma. Tornadoes appear most frequently in the U.S. during the months of April, May, and June.

5.3.5. Lightning

In addition to being a personal hazard, lightning strikes can ignite combustible building materials and storage. Lighting strikes can also damage electrical equipment and wiring. According to the NFPA, it causes more deaths and property damage than floods, hurricanes, and tornadoes combined. Lightning strikes are usually associated with the occurrence of severe thunderstorms. NFPA 780, "Installation of Lightning Protection Systems," contains a risk assessment guide that is in the appendix of this standard. This guide is intended to assess the risk of loss from lightning.

5.3.6. Arctic Freeze

An arctic freeze is a term applied to a condition where unusual freezing conditions occur during the winter months. The areas of the U.S. that are susceptible to this phenomena are those that are not normally subject to freezing temperatures. This results from a temporary change in the weather pattern by which a frigid air mass moves further south than normally expected. A temperature drop of 15°F to 20°F (8.3°C to 11.1°C) can be expected and may last from several days to weeks. Forewarning of this type of event is available from the National Weather Service. Broken pipes, including domestic water and fire protection, are the most common result of an arctic freeze. Upon thawing, broken pipes can cause significant water damage in a warehouse. Additionally, utilities may curtail gas supplies to industrial customers, preferring to supply residential customers.

Safely operating a warehouse in an area prone to natural perils will require certain provisions to mitigate the effects of the event. The warehouse should be constructed using features designed to withstand the most severe event. In particular this should include fire protection and other emergency equipment, such as emergency lighting. Despite these precautions, it should be recognized that significant damage may still result to the warehouse building and contents. A emergency plan, as outlined in Chapter 9, should therefore be developed based on the potential loss scenarios.

5.4. Exposures from Surrounding Activities

5.4.1. Adjacent Facilities, Airports, Highways, and Railroads

The exposure potential from adjacent facilities due to catastrophic incidents such as a fire, explosion, or a chemical release should be identified when conducting a site assessment. Reassessments of nearby facilities may also be warranted as changes to the surrounding exposures occur over time. High hazard or poorly protected operations may present a risk to a nearby chemical warehouse. The most effective approach for minimizing the exposure is usually sufficient spatial separation or fire walls. Locating a chemical warehouse adjacent to airports, highways and railroad lines may also result in an exposure, albeit remote.

5.4.2. High Pressure Flammable Gas and Liquid Transmission Lines

A network of underground pipelines is used to transport vast quantities of flammable gases and liquids throughout the United States. These pipelines traverse long distances and operate under high pressures.. Large leaks from a high pressure flammable gas or liquid transmission line, although rare, have caused major fires and explosions.

A single reference detailing the location and type of all pipelines does not currently exist. Information concerning pipelines should be obtained on a local level. This can include the local government officials or the owner/operators of these pipelines.

5.4.3. Riot and Civil Commotion

The occurrence of riot and civil commotion involving the general population is most likely to occur in high crime and economically depressed areas, although labor unrest can also precipitate such incidents. In case of a widespread riot, response from police and fire services may be delayed or unavailable. Past experience has shown that, in incidents involving the general population, facilities storing nonconsumer items are not likely to be targeted by looters. Additionally, facilities constructed from noncombustible materials and having a sprinkler system are less likely to be damaged by a fire that may occur. Mitigative features, as presented in this guideline, may also be useful in reducing the facility's vulnerability.

5.5. Emergency Responders

Accessibility to emergency responders during an emergency should be part of the site assessment. The travel routes of emergency responders should be evaluated taking into account travel distances and possible obstructions such as railroad crossings. When estimating the response time to the site, the time from when the request for help is received until application of fire water should be used. Emergency responders should have complete access to the perimeter of the building, which may require additional access roads. Fire hydrants should also be strategically placed near and around the facility so that an adequate and reliable water supply is available.

Assessment of the adequacy of emergency response should be based on firefighting capability and the ability to handle hazardous material incidents. The development of a response plan with emergency responders and, as appropriate, emergency management officials, will enable emergency responders to arrive better prepared to deal with an emergency.

5.6. Adequacy and Reliability of Public Utilities

The adequacy and reliability of the public utilities should be reviewed as part of a site assessment. Water supplies in particular should be evaluated against fire water requirements. If the water supply is not adequate or sufficiently reliable, an on-site supply, such as a fire pump and tank, may be needed.

References

Building Officials & Code Administrators International, Inc., "The BOCA National Building Code," Country Club Hills, IL, 1993.
Federal Emergency Management Agency, "Guide To Flood Insurance Rate Maps," FIA-14, Baltimore, MD, May, 1988.
Industrial Risk Insurers, "Arctic Freeze," IM.15.5.1, Hartford, CT, March 1, 1996.
Industrial Risk Insurers, "Earthquake," IM.15.2, Hartford, CT, June 3, 1996.
Industrial Risk Insurers, "Flood," IM.15.4, Hartford, CT, December 1, 1995.
Industrial Risk Insurers, "Riot and Civil Commotion," IM.1.11.1, Hartford, CT, June 1, 1993.
Industrial Risk Insurers, "Tornadoes," IM.15.1.2, Hartford, CT, March 1, 1996.
Industrial Risk Insurers, "Windstorms," IM.2.0.1, Hartford, CT, February 2, 1990.
International Conference of Building Officials, "Uniform Building Code," Whittier, CA, 1991.
National Fire Protection Association, "Standard for the Installation of Lightning Protection Systems," NFPA 780, Quincy, MA, 1995.

Additional Reading 65

Southern Building Code Congress International, Inc., "Standard Building Code," Birmingham, AL, 1991.
Department of Energy, Federal Energy Regulatory Commission, "Major Natural Gas Pipelines," Washington, D.C., September 30, 1993.

Additional Reading

Factory Mutual Engineering Corporation, "Evaluation of Flood Exposure," Loss Prevention Data 9-13, Norwood, MA, February, 1993.
Factory Mutual Engineering Corporation, "Maps of Earthquake Zones," Loss Prevention Data 1-2S, Norwood, MA, February, 1987.
Factory Mutual Engineering Corporation, "Protection Against Fire Exposure (from Buildings or Yard Storage)," Loss Prevention Data 1-20, Norwood, MA, March, 1979.
Industrial Risk Insurers, "Protection of Buildings from Exterior Fire Exposures NFPA 80A-1987," IM.2.0.5, Hartford, CT, December 2, 1991.
Industrial Risk Insurers, "Windstorm Data Maps," IM.2.0.1.1.A, Hartford, CT, December 2, 1991.
National Fire Protection Association, "Recommended Practice for Protection of Buildings from Exterior Fire Exposures," NFPA 80A, Quincy, MA, 1996.
Pielke, R. A., *The Hurricane*. Routledge, New York, 1990.

6 Design and Construction

6.1. Synopsis

The design and construction of a chemical warehouse should consider possible loss scenarios that may affect employees, the surrounding population, the environment, the warehouse building and business continuity. Protection and mitigation features should be consistent with the characteristics of the materials stored, environmental and population vulnerability, and potential natural perils.

6.2. Construction Documents—Approvals and Permits

A chemical warehouse facility should be designed by a project team knowledgeable in the hazards presented by the materials stored and the special building features which may be required. Life safety, community exposure, containment features, preservation of property (including interruption of business, lost inventory, etc.) and geological factors should be considered. (See also Chapters 3 and 5 for information involving regulations and facility siting.)

The hazard classifications and methods of protection of various materials identified in this Guideline may not be consistent with the classifications, protection methods or mitigation means found in the applicable building construction or fire codes. Compliance with building codes will not always ensure an adequate warehouse design. They may, in fact, provide only minimum protection. The appropriate level of protection beyond that required by code or regulation is a business decision.

Information required in submittal or construction documents is delineated in most building codes. Table 6-1 identifies the three major organizations in the U.S. that promulgate design codes and/or standards.

TABLE 6-1
Model Code Organizations and Promulgated Primary Codes

Building Officials and Code Administrators International, Inc. (BOCA) Country Club Hills, Illinois
- National Building Code
- National Fire Prevention Code
- National Mechanical Code
- National Plumbing Code

International Conference of Building Officials (ICBO) Whittier, California
- Uniform Building Code
- Uniform Fire Code*
- Uniform Mechanical Code
- Uniform Plumbing Code**

Southern Building Code Congress International, Inc. (SBCCI) Birmingham, Alabama
- Standard Building Code
- Standard Fire Prevention Code
- Standard Mechanical Code
- Standard Plumbing Code
- Standard Gas Code

* Published by the International Fire Code Institute and endorsed by ICBO.
** Published by the International Association of Plumbing and Mechanical Officials.

(The table is not intended to be all inclusive with respect to the codes published by these organizations.) Figure 6-1 shows the general areas of code influence in the United States.

In addition to the organizations and codes listed in Table 6-1, the National Fire Protection Association promulgates standards and codes that address fire protection and life safety in building construction. Two of the NFPA's most frequently adopted codes are NFPA 70, "National Electrical Code" and NFPA 101, "Code for Safety to Life from Fire in Buildings and Structures," more commonly known as the Life Safety Code. Both of these codes are adopted as standards of the American National Standards Institute (ANSI).

Efforts are underway within the model code organizations to develop a single set of national codes for adoption and use throughout the U.S. These organizations have formed the International Code Council (ICC) to jointly develop these documents. The ICC has already published a

6.2. Construction Documents—Approvals and Permits

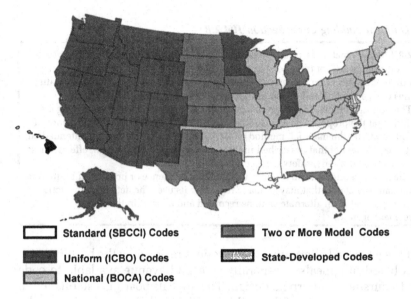

FIGURE 6-1. *General Areas of Building Code Influences*

mechanical and plumbing code and has targeted the year 2000 for release of building and fire codes. The NFPA has also joined the ICC effort in development of a single fire code.

County governments and states, in particular, have agencies or entities other than local building and fire departments that are involved in the approval process. Local processes may involve zoning approvals, and special permits for land use, utilities, storage operations, hazardous uses, etc.

For a chemical warehouse, special occupancy requirements are associated with facilities classified as a Hazardous Use. All three model codes allow the building official to require a technical opinion or report identifying and developing alternate methods of construction or protection from hazards that are presented by the storage of hazardous material. The technical evaluation or report is helpful in demonstrating that the proposed building design features, storage arrangements, materials, etc. meet the intent of the code. Table 6-2 provides an example of code language for alternative designs.

Chemical warehouses storing materials that are not classified as a Hazardous Use will be subject to provisions for a less restrictive Storage Use.

TABLE 6-2
1994 Uniform Building Code, Section 104.2.8

> **104.2.8** Alternate materials, alternate design and methods of construction. The provisions of this code are not intended to prevent the use of any material, alternate design or method of construction not specifically prescribed by this code, provided any alternate has been approved and its use authorized by the building official.
>
> The building official may approve any such alternate, provided the building official finds that the proposed design is satisfactory and complies with the provisions of this code and that the material, method or work offered is, for the purpose intended, at least the equivalent of that prescribed in this code in suitability, strength, effectiveness, fire resistance, durability, safety and sanitation.
>
> The building official shall require that sufficient evidence or proof be submitted to substantiate any claims that may be made regarding its use. The details of any action granting approval of an alternate shall be recorded and entered in the files of the code enforcement agency.

If the proposed alternative to the code is rejected by the building official, a board of appeals is generally a second avenue available to code users to pursue the alternate design. The appeals board is established to hear and decide appeals of the building official's decisions or orders relative to the application and interpretations of the code. The board may not waive requirements of the code.

6.3. Means of Egress

Life safety in a chemical warehouse is best assured through control and mitigation of hazards. However, emergency evacuation remains an essential element in warehouse life safety.

The means of egress from a chemical warehouse consists of those components that are necessary to provide a safe path of exiting for warehouse and non-storage area occupants. Due to its critical function, the safe path needs to be available for use under normal and emergency conditions.

Design considerations for the means of egress in a chemical warehouse include:

- Storage arrangement and aisle spacing.
- Occupant load and distribution.
- Building design, layout or internal partitioning.
- Hazard of stored chemicals.
- Presence of automatic fire suppression systems.
- Presence of fire detection and alarm systems.

6.3. Means of Egress

- Exit discharge and access to a safe area away from the warehouse.
- Building security.

A chemical warehouse may contain multiple uses or occupancies that should be evaluated separately for means of egress design. Personnel facilities (offices, breakrooms, lavatories, etc.) should not be located in the same fire area as warehouse operations. At least a one-hour fire-rated separation between warehouse and non-storage areas should be provided; a greater fire rating could be required.

Specific criteria and requirements for means of egress design can be found in local and national building codes as well as in the NFPA 101, Life Safety Code. The Life Safety Code is one of the most widely used standards for means of egress design in the U.S. Means of egress requirements typically entail the following elements:

- Definitions of an exit.
- Arrangement/location of exits.
- Capacity of exits.
- Travel distance to an exit.
- Enclosure/separation of the means of egress.
- General and emergency lighting.
- Exit signage and marking the egress path.

When evaluating the types of exits to be provided for the chemical warehouse, the designer should consider the following:

- Elevation differences between grade and the warehouse floor slab at exterior exit locations.
- Provision of adequate exit stairs or ramps at warehouse truck docks (especially depressed docks).
- Provision of pedestrian doors adjacent to overhead roll-down doors or horizontal sliding doors.
- Use of enclosed exit stairs to serve multiple level areas (i.e., two story office space adjacent to or within a single story warehouse).
- Use of inverter or generator back-up power for emergency lighting systems (especially for large area warehouses, where battery-powered "bug-eye" lights are not practical).
- Use of open or enclosed exit stairs to serve mezzanine areas of the warehouse.
- Provision of adequate slopes for ramps serving spaces or rooms employing curbing such as for containment.
- Use of delayed egress locks on exit doors that need to serve as a security barrier.

- Use of horizontal exits in fire-rated walls that may be required for warehouse subdivision.
- Use of vehicle ramps as possible egress paths for warehouse occupants.
- Provision of panic-type hardware on doors at locations requiring rapid evacuation under emergency conditions, such as from a flammable liquid storage room.
- Timing required for employees to reach exits under adverse conditions.

Exits from the warehouse need to be arranged so that a single emergency condition, such as fire or accidental spill, will not block access to all exits. Exits from non-storage areas (such as locker rooms and offices) should provide a direct or protected path to the exterior and not rely solely upon travel through the warehouse to egress the building. Minimum provisions in the Life Safety Code and other codes address conditions where multiple exits are required, the remoteness of these exits from one another, paths of travel and time required to reach these exits, emergency lighting of exit pathways, and the need for fire-rated enclosure of the exit path.

6.3.1. Travel Distance

Different travel distances are established for unsprinklered and sprinklered buildings with the exception of certain occupancies classified as Hazardous Uses. Maximum allowable travel distances for individual occupancies should be determined by reference to the applicable code.

Warehouse occupancies classified as Hazardous or High Hazard are generally required to have a shorter travel distance to an exit than most other occupancies. The maximum exit travel distance in a hazardous occupancy is typically 75 feet, though some codes and "alternate method" considerations permit greater distances. In comparison, the travel distance to an exit in a warehouse classified as an Ordinary Hazard storage occupancy by the Life Safety Code can be as long as 200 feet for an unsprinklered building. This travel distance can be increased to 400 feet in a fully sprinklered building.

In those cases where an engineering evaluation is used to establish life safety criteria, travel distance to an exit should at a minimum be determined by an evaluation of occupancy hazards, occupant loading, egress pathways and hazard protection/mitigation features of the warehouse building.

6.4. Environmental Protection

Means of egress criteria may also rely upon the presence of other building features in the chemical warehouse. Examples include automatic sprinkler systems, fire detection/warning systems (optical flame-sensing, smoke, heat, etc.) and smoke removal systems. In certain situations, these systems may permit less restrictive arrangements of the means of egress components.

As part of overall emergency preparedness, warehouse operators should conduct regular drills to ensure occupant familiarity with building exits and the paths of exit travel. These drills establish a routine which provides a higher degree of life safety for building occupants.

6.4. Environmental Protection

Chemical warehouses should be constructed with provisions for spill containment and drainage to provide protection to the surrounding property, ground water, surface water and public sewers in the event of an accidental chemical release or discharge of sprinkler water. This protection should be commensurate with the toxicity of the stored materials and the risk and consequence of exposures.

When designing a warehouse floor, consideration should be given to preparation of the site upon which the warehouse is built; the type and grade of the concrete, and the techniques used to install it; the preparation and treatment of the floor's surface; and maintenance and repair. The level of protection applied to any given warehouse design should be consistent with the hazards associated with the planned occupancy and the level of risk with which the warehouse owner is comfortable. An impervious warehouse floor is not always justified. However, since environmental exposures often carry serious consequences, these measures should be carefully considered for their applicability to any particular warehouse. A risk analysis which considers the nature of the chemicals stored, fire risk, groundwater sources and geology can be a useful tool in making decisions regarding such design features.

6.4.1. Containment and Drainage Capacity Considerations

Where containment and drainage options are deemed appropriate, the capacity of primary containment or a drainage system discharging into secondary containment should account for the maximum probable release of chemicals and fire protection water. Depending on the size and type of the commodity involved, there exists the potential for a signifi-

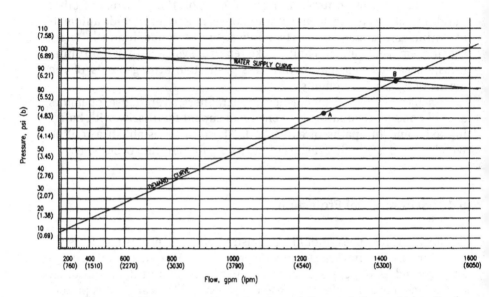

FIGURE 6-2. **Sprinkler System Water Supply and Demand Curves Using Semi-exponential ($N^{1.85}$) Graph Paper. Point A: Design Flow Rate and Pressure. Point B: Actual Flow Rate and Pressure.**

cant volume of material to be released, particularly under fire conditions. The volume of contaminated sprinkler or hose stream water runoff will most likely greatly exceed that of a potential chemical release. The anticipated sprinkler and hose stream discharge should also be factored into the containment and drainage design.

Though standards may not require it, the actual sprinkler discharge flow rate should be considered rather than the design flow rate, which may be lower. The actual flow rate can be determined by plotting the sprinkler system water supply and demand curves (see Figure 6-2). The intersection of the water supply and demand curve will be the actual flow rate. This estimate is contingent upon operating the number of design sprinklers and the anticipated number of hose streams.

Sprinkler system flows are typically calculated to supply water for a duration of 1 or 2 hours. Therefore, a sprinkler system having a combined discharge of 2500 gpm (9500 lpm) over a 1 hour period would produce 150,000 gallons (570,000 liters) of water. If this occurred in a warehouse having a 50,000 ft.^{2} (4,650 m^2) containment area, the depth of water would be 4.8 in. (12.2 cm). (See Table 6-3)

6.4. Environmental Protection

TABLE 6-3
Required Containment Depth (in.(cm)) for 1 Hour of Fire Protection Waterflow

Containment Area ft² (m²)	1,000 gpm (3800 lpm)[1]	2,500 gpm (9,500 lpm)[2]	4,000 gpm (15,200 lpm)[3]
5,000 (465)	19.3 (49)	48.3 (123)	77.2 (196)
10,000 (930)	9.6 (24.4)	24 (61)	38.4 (97.6)
20,000 (1,860)	4.8 (12.2)	12 (30.5)	19.2 (48.8)
50,000 (4,650)	1.9 (4.8)	4.8 (12.2)	7.6 (19.2)
100,000 (9,300)	.97 (2.4)	2.4 (6.1)	3.9 (9.6)
200,000 (18,600)	.48 (1.2)	1.2 (3.1)	1.9 (4.8)

[1] 60,000 gal (228,000 l) total volume
[2] 150,000 gal (570,000 l) total volume
[3] 240,000 gal (912,000 l) total volume

A warehouse utilizing a fire suppression or fire extinguishment system designed to minimize the number of operating sprinkler heads, maximize the flow rate per head and obtain suppression or extinguishment quickly (such as an early-suppression fast-response (ESFR) or a closed-head foam-water sprinkler system) may only require a duration of 30 minutes. However, if a system that provides fire control rather than suppression or extinguishment is used, a longer duration (1-2 hours) may be appropriate. Consideration should also be given to the additional fire water from fire hose streams due to manual fire fighting efforts. For further explanation of the terms "fire suppression," "fire control," and "fire extinguishment," refer to Chapter 7. Fire protection systems can be utilized, subject to their limitations, that do not generate large quantities of water. These systems include dry chemical, gaseous agents, and high expansion foam.

The design parameters for fire protection systems, which are covered in more detail in Chapter 7, include criteria useful in estimating the potential volume of runoff generated by water-based sprinkler systems.

Rainfall may also contribute to the volume requirements with outdoor storage and open secondary containment. The rainfall accumulation could be based upon the highest intensity expected over a 24 hour period. Twenty-five year rainfall records provide a useful history for estimating purposes.

A strategy of employing only primary containment may be more desirable with existing facilities or in a new building where it is not nec-

6. Design and Construction

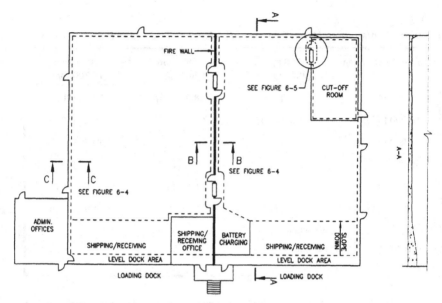

FIGURE 6-3. *Chamical Warehouse Showing Containment Features*

essary or practical to install a drainage system, or where operations issues or the low hazard nature of the materials stored render a primary containment strategy appropriate. Primary containment may consist of a liquid-tight depressed warehouse floor. Alternatively, the warehouse can be equipped with curbs or ramps. Figures 6-3, 6-4, and 6-5 show details of a depressed floor, and curbs and ramps.

Secondary containment can be accomplished by a number of methods, among them, exterior containment or a depressed loading dock area.

A drainage system could utilize wall scuppers, underground floor drains, and/or trenches designed to lead to a safe location. The liquid-tight floor should also be pitched toward the drainage devices. The system should provide sufficient capacity to prevent overflow into unwanted areas. It should also take into account the liquid height limitation in the primary containment area. Some over-capacity should be designed into the system to accommodate the potential for debris-clogged drains. The drainage system should also incorporate design features which are intended to prevent flame spread through trenches.

Figure 6-6 shows a representative floor plan for a generic chemical warehouse with two possible containment options. The left side of the warehouse drawing (labeled Option A) illustrates a trench and piping

6.4. Environmental Protection

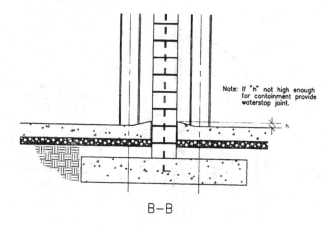

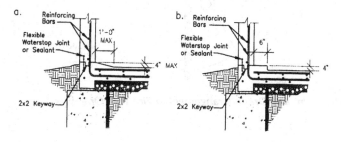

FIGURE 6-4.
B-B: Floor Containment at Fire Wall.
C-C: Floor Containment at Outer Wall.

option to collect and transport fire water to a remote exterior containment area (also shown in Figure 6-7). The second option, illustrated on the right side of the warehouse plan as Option B (also shown in Figure 6-8) utilizes a ramped up floor slab at the floor perimeter for primary containment. Secondary containment utilizes the outside recessed loading dock area. Run-off is directed to the loading dock area through the loading dock door. Note that the shipping and receiving area on both options slopes down, away from the dock doors, in order to maximize the primary containment of the warehouse floor. One possible configuration is to slope from a high point about 20 feet from the dock doors, with pitch from the high-point toward the doors, and high-point toward the general ware-

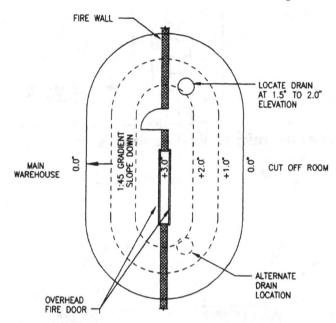

FIGURE 6-5. *Ramp and Drain Arrangement at Fire Wall Openings*

house areas. This allows the shipping and receiving area, which is a likely area for leaks to occur, to drain away from the warehouse proper instead of toward it.

The presence of a trench/pipe system in Option B, used to remove rain accumulations, increases the risk that contaminated fire water might be inadvertently discharged into the sanitary sewer system. In order to reduce this risk, it is necessary to install a motorized valve in the drainage piping. This valve could be arranged to close automatically in the event of a fire through a signal from the fire alarm panel and manually as needed. It may also be desirable to use a valve which will automatically close in the event of loss of power. Alternatively, this valve could be normally closed and opened when needed.

Contaminated fire water should never be routed to the sanitary sewer. Where a combined sewer system makes separate discharges impossible, it is critical that the valve be closed in the event of fire water discharge. Under no circumstances should the contaminated fire water be allowed to discharge to the local publicly owned treatment works (POTW). A trap tank or sump to separate any light layer may be appropriate, particularly where contaminants are not water-miscible.

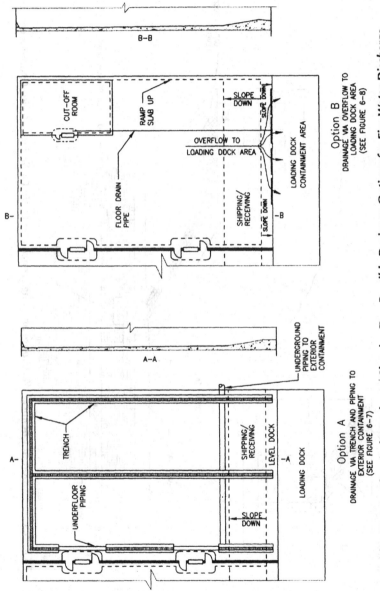

FIGURE 6-6. *Chemical Warehouse Showing Two Possible Drainage Options for Fire Water Discharge*

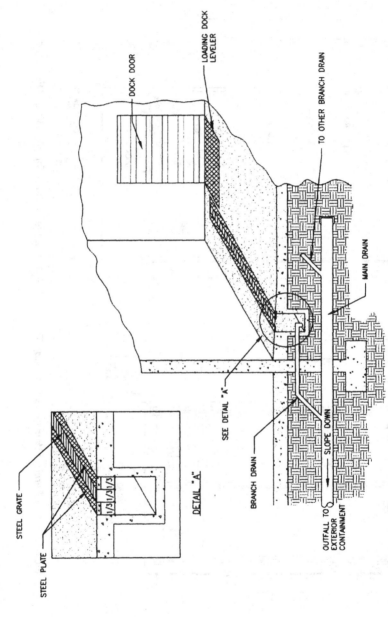

FIGURE 6-7. *Option A: Chemical Warehouse Showing Trench and Drain Option to Exterior Containment*

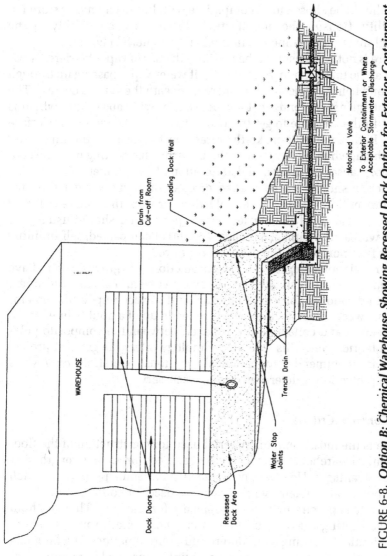

FIGURE 6-8. *Option B: Chemical Warehouse Showing Recessed Dock Option for Exterior Containment*

6.4.2. Warehouse Floor System

In preparing the site, a subbase of granular material is placed over a graded and compacted subgrade. If a vapor barrier is to be used, a 6.0 mil polyfilm seamless sheet should be used. The composition of the vapor barrier should be chosen to be compatible with liquid chemicals stored in the facility. Consult the manufacturer to assure compatibility of the chemical inventory planned with the barrier material chosen.

There is not consensus on the merits of the use of vapor barriers. Those in favor argue that vapor barriers can limit water vapor passing up through the soil and being trapped in the concrete beneath the surface coating. The argument against contends that a vapor barrier will cause water which is resident in the concrete to get trapped between the concrete and surface coating. Both arguments make their case on the basis of limiting water vapor which, when trapped under the impermeable coating over the concrete floor, causes cracking and deterioration of the concrete.

Concrete should not be poured directly on a vapor barrier as this would prevent drainage of excess bleed water from the concrete during curing, and tearing of the barrier. The vapor barrier should usually be covered with a minimum of 3 inches (75 mm) of compacted, self-draining granular fill upon which the concrete is poured.

As regulations regarding the protection of groundwater have increased, more sophisticated lining systems such as geotextile liners have been developed for use in landfills and surface impoundments for hazardous waste. Such technologies are made up of a double liner and a leak detection and collection layer placed between the composite polymer and fabric layer. This level of technology offers the greatest protection, but is recommended for chemical warehouse operations only when absolute protection is determined to be necessary.

6.4.3. Concrete Criteria

Concrete is the most common material used for construction of the floor of a chemical warehouse. Its relative cost and compression strength offer obvious advantages. However, concrete also has some properties which present potential problems when floors are used for containment.

Concrete is porous and has a propensity for cracking. Thus, without protective coatings, certain chemicals can penetrate through the concrete, potentially leading to underground contamination. Cracking can allow corrosive chemicals to reach the reinforcing steel bars which are sometimes embedded in the concrete. Rusting and deterioration of the

6.4. Environmental Protection

concrete can then result. Further, cracks allow chemical spills to move below the concrete and contaminate the underlying soils.

Use of high quality concrete can help resist deterioration. Techniques used to install concrete are also important to the success of any containment strategy.

Concrete specifications for facilities which handle agriculture chemicals are provided in "Designing Facilities for Pesticide and Fertilizer Containment" available from the Midwest Plan Service. Specifications in this document provide a conservative approach to the protection of a chemical warehouse.

The surface of the floor should be smooth with the exception of the joints formed when slabs of concrete are poured. Expansion joints allow the concrete to expand, contract and move without cracking. Expansion joints should be inserted into flat concrete surfaces and should not intersect doorways, ramps for lift truck travel, rack supports, or within 10 feet of the intersection of walls, doors or corners. Construction joints may be formed when a concrete slab is poured and allowed to dry for more than one-half hour before a second is poured—flush against the first. Control joints are intended to provide a preestablished seam where the concrete will crack. Constructing such seams recognizes that concrete cracking is inevitable and identifies the point at which the cracking will occur. Controlled cracking allows the warehouse designer to control stresses under that point. If concrete were allowed to randomly crack, the cracks would be more difficult to repair and concrete would be likely to further deteriorate and spills would be more difficult to contain or clean up before entering cracks.

Joints must then be sealed with an elastomeric sealer in order to prevent seepage through the seam or crack. To seal control joints and cracks, a chemically resistant joint sealer is applied in the seam joint. There are many different sealant materials and specifications for installation which should be considered at the design stage.

A waterstop, which is installed before pouring the concrete, may be effective in sealing joints, but is not applicable for cracks. A waterstop is used as a barrier to water leakage under floor joints, where curbing is used, or at wall/floor junctions. Joints are then made above the waterstop and sealed. The waterstop can provide a second seal if the sealant in the joint fails (Figure 6-4).

6.4.4. Surface Preparation

Surface preparation, coating application methods and coating materials chosen or specified can effectively help prevent potential damage to concrete and from migration of chemicals through concrete. Proper surface preparation yields the best adhesion of the coating to the concrete. The failure of coatings is most often due to the improper preparation of the concrete surface.

Coatings failure occurs by delamination (separating by layers) or by peeling (separating from the concrete). The latter happens most frequently when the adhesive bond with the concrete is lost. This type of failure occurs when coatings are applied without first removing the surface contaminants and laitance. Laitance, a film which can form on the surface of the concrete, is unreacted cement or cement which has risen to the surface of the concrete surface due to over-working fresh concrete.

Decisions regarding the method of preparation of concrete surfaces are complex and the application of surface coatings to a concrete floor should be made after carefully analyzing the geography of the site, the chemicals to be stored, the type of cement used, the potential exposure conditions and the life expectancy.

6.4.5. Coatings and Sealers

Coatings should be applied in such a way as to provide a continuous film of protection that has a relatively constant thickness and adheres strongly to the substrate. Consult the manufacturer for application requirements. The number of coats varies with the formulation, the application system and the requirements of the operation. The thickness of each application depends on the specification for the coating and the chemicals to be stored.

Concrete surface temperature, ambient temperature, and relative humidity can affect the performance of the coating. If the thermal expansion coefficients of the coating and concrete are widely different, separation at the interface is likely.

When appropriate materials are properly applied, coatings and sealers can effectively protect the concrete substrate. Decisions regarding the materials used to coat concrete substrates should be influenced by the chemicals which will be stored, and how particular coating materials react with those chemicals in accidental spill conditions. Also important in the process is the tendency of the particular concrete substrate chosen to shrink and crack.

In most chemical occupancies, the preparation of floors and the coating formulation to be used are not straightforward decisions. Research which has been conducted on the effectiveness of concrete coatings and offers more information on these issues than do current government

6.4. Environmental Protection

regulations. Recent tests at TVA Environmental Research Center illustrate these issues. Information on this research is covered in the *Journal of Protective Coatings and Linings*, July 1995, "Protective Secondary Containment from Pesticides and Fertilizer."

Coating applications will vary in terms of weight and volume gain and loss, surface embrittlement, softening, blistering and color change, depending upon the nature of the chemicals to which they are exposed.

Chemical compatibility issues which affect the process of choosing appropriate coating formulations for chemical warehouses are very complex. In addition to the coating method and materials, the number and thickness of applications may be influenced by particularly aggressive chemical environments. Therefore, consultation with the coatings manufacturer's engineers is recommended.

6.4.6. Maintenance and Repair of the Floor

While most concrete defects occur as the result of flaws in design and installation, even good designs sometimes experience defects, earthquakes or similar events, freeze/thaw cycles, vapor/moisture permeability or overloading which lead to a need for repair. The floor should be checked periodically for cracks, spalling, bug holes, honeycombs, or other flaws. Where such flaws appear, they should be repaired and sealed.

Cracks are either dormant or active. Dormant cracks can be repaired without experiencing additional cracking. Active or working cracks will most often crack again, even when repaired. The method and material for repair can influence the effectiveness of the repair.

As with construction factors, repair materials should be chosen to match the particular needs of the situation. Most repair materials are either cement based or synthetic resin based.

Since repairs are made on concrete which has already dried, additional shrinkage will not occur. The material used in repair should also be either shrink-free or have the capacity to shrink without losing bond.

However, the repair material is subject to the same stresses as the concrete with regard to temperature variation. Ultimately, the repair will prove most durable if the repair material has a similar coefficient of thermal expansion as the original material.

A variety of techniques are employed in concrete repair, depending on the nature of the crack and the material chosen. Concrete repair professionals may be hired for this task, or instructions from the manufacturer of the repair material should be followed.

6.4.7. Airborne Effluent

Control of airborne effluent such as gases, vapors and particulate matter in smoke is very difficult in a chemical warehouse environment. Design criteria currently does not exist for systems to mitigate airborne releases resulting from fires, decomposition or incompatible chemical reactions. Therefore, the emphasis should be on preventing these incidents from occurring.

6.5. Fire Mitigation Construction Features

The building materials used in the construction of a chemical warehouse can be based on a variety of factors such as building size and geographic location, material costs, energy efficiency, occupancy, construction codes, insurance criteria, owner's investment strategy (owned vs. leased building) and building aesthetics. The building materials, when combined in a structure, can be classified in accordance with their fire resistive performance.

Five basic construction classifications identified in U.S. building codes and NFPA standards are described in Table 6-4.

Regardless of construction type, building construction generally falls into one of three categories: fire resistive, noncombustible or combustible. These categories are described in Table 6-5 along with a reference to a construction classification.

Combustible materials of one form or another can be found in roof assemblies, exterior wall assemblies or insulation, insulated wall or ceiling assemblies for use in controlled temperature situations (i.e., coolers, freezers, etc.), interior partitions, ceiling assemblies, and interior finish materials for floors, walls, ceilings, and trim (i.e., carpeting, wallpaper, paneling, etc.).

Generally, chemical warehouses whose contents classify them as hazardous occupancies should be constructed of fire resistive or noncombustible construction. Fire resistive construction should consist of materials that will withstand the anticipated fire exposure of a given duration without structural failure. The following standards and test methods are commonly used to establish selected fire resistance ratings:

- ASTM E119, "Standard Methods of Fire Tests of Building Construction and Materials"
- UL 1709, "Rapid Rise Fire Tests of Protection Materials for Structural Steel"

6.5. Fire Mitigation Construction Features

TABLE 6-4
U.S. Building Construction Classifications

Construction Type	Basic Description
I	• Fire resistive and noncombustible construction • Subclassified into two categories where structural frame carries a fire resistance rating of 3 or 4 hours • Floor construction is 2 or 3 hour rated • Roof construction is 1½ or 2 hour rated
II	• Noncombustible construction • Subclassified into three categories where structural frame carries a fire resistance rating of 2 hour, 1 hour or 0 hour (unprotected noncombustible construction) • Floor construction is 2, 1 or 0 hour rated • Roof construction is 1 or 0 hour rated
III	• Primarily combustible construction • Subclassified into two categories where columns and interior bearing walls are 1 or 0 hour rated • Exterior bearing wall construction is noncombustible and 2 hour rated • Floor and roof construction is 1 or 0 hour rated
IV	• Combustible construction commonly referred to as heavy timber, mill or plank-on-timber • Exterior and interior bearing walls are of noncombustible construction and 2 hour rated • Interior structural members, floors and roofs are solid or laminated wood and meet minimum dimensional criteria
V	• Wood or other approved combustible construction • Subclassified into two categories where all structural elements (e.g. columns, floor, roof, etc.) are either 1 or 0 hour rated

- ASTM E1529, "Standard Test Method for Determining Effects of Large Hydrocarbon Pool Fires on Structural Members and Assemblies"

Fire resistive construction is typically tested and listed in accordance with ASTM E119. This test method is suitable for fire resistive needs in most warehouses whose contents can be classified as ordinary hazard.

TABLE 6-5
Building Construction Categories

Construction Category	Basic Description	Example of Classification[1]
Fire Resistive	Construction utilizes materials that will withstand a specified fire exposure for a defined period of time. This time period is established either by a standard fire test procedure or recognized calculation methods derived from test data and experience.	Type I (433) Type I (332) Type II (222) Type II (111)
Noncombustible	Construction utilizes materials that do not contribute to or propagate fire. These materials may be tested in accordance with a recognized fire test standard when their "noncombustible" characteristics are not readily evident. Noncombustible materials may lose their strength under conditions of fire or heat exposure.	Type II (000)
Combustible	Construction consists of materials which will ignite and burn. Combining combustible materials with otherwise noncombustible materials may result in an overall combustible composite assembly.	Type III (211) Type III (200) Type IV (2HH) Type V (111) Type V (000)

[1]The example identifies a U.S. construction classification based upon NFPA 220, "Types of Building Construction." The Arabic numbers following each type of construction (e.g. Type I, II, etc.) indicate fire resistance ratings for certain structural elements. The first number addresses exterior bearing walls. The second number addresses the structural frame when supporting more than one floor. The third number addresses floor construction. Other construction codes have similar and more detailed classifications.

UL 1709 and ASTM E1529 tests use a rapid rise fire growth curve that is typical of a large hydrocarbon pool fire versus a slower steady growth fire curve in ASTM E119. Fire resistive construction tested and listed in accordance with UL 1709 and ASTM E1529 should be considered for use in warehouses whose contents include flammable liquids or chemicals with a high heat release potential under fire situations. Alternatively, the realities of intense fires may warrant reconsideration of roof fireproofing materials, with greater emphasis on the fire-resistivity of walls or building separation features.

6.5. Fire Mitigation Construction Features

6.5.1. Fire-Rated Separations

When designing for fire safety in a chemical warehouse, building area fire separations may be needed to:

- separate personnel facilities from warehouse areas.
- limit the risk or size of a potential property, business interruption, contents or environmental loss
- segregate different stored materials due to their hazardous properties
- increase the likelihood of control in a fire situation by exposing a limited area that is considered manageable by automatic and manual suppression efforts, or by a building's containment/drainage design
- meet building or fire code requirements for building area limitations, separation of occupancies/hazards, travel distance considerations, or other provisions

Building area fire separations primarily consist of fire resistive rated walls and in some cases floors, ceilings and roofs. In a single story chemical warehouse, the prevalent component used as a fire separation is a wall with a fire resistance rating and suitably protected openings. Although fire separations are categorized or classified differently by a variety of entities, there are two basic types of fire separations: fire walls and fire-rated barriers/partitions.

Fire-rated barriers/partitions are those walls that are supported in some manner by the building's structural frame, floor, ceiling or roof. These walls rely upon the integrity of the supporting construction to remain in place and to be able to withstand a given fire duration.

Fire walls are those walls that are designed to be either independent of the building's structural frame and major building components, or designed with sufficient stability and strength to remain standing in the event of building collapse on one or the other side of the wall. These types of walls are typically constructed of masonry or concrete materials.

Table 6-6 provides an example of one insurance carrier's classification method and terminology for fire separation walls. This example demonstrates the range of required fire-rated separations. These categories relate to the ability and the length of time each can withstand a standardized fire exposure, the capacity to maintain continuity and stability during periods of prolonged and excessive heat, and the physical forces which occur to each during a fire (such as collapse).

TABLE 6-6
Rating of Fire Walls, Barriers, and Partitions

Type of Wall	Fire Rating	Configuration
Standard fire wall	4-hour minimum blank with no openings	Parapet extends above roof with wingwalls, endwalls or extensions
Fire wall	3 to 4 hour with protected openings	Parapet extends above roof with wingwalls, endwalls, or extensions
Fire barrier	2 to 3 hour with protected openings	Wall extends from floor to beneath roof deck
Fire partition	1 to 2 hour with protected openings	Wall extends from floor to ceiling

Source: Industrial Risk Insurers

Table 6-7 identifies general fire wall construction features.

Fire barriers, described in Table 6-8, should be used to separate special hazards within those warehouses that require segregation from the general storage. Flammable liquids, organic peroxide formulations and aerosols are examples of materials which may require this special segregation.

Fire Partitions are similar to *Fire Barriers* in every respect except they have a 1 to 2 hour fire rating. Fire partitions are generally used to further subdivide areas and can also be supported by fire resistance rated columns, floors, and ceilings.

Fire resistance rated wall assemblies are tested and listed by nationally recognized testing laboratories such as Underwriter's Laboratories, Warnock-Hersey or Factory Mutual. These listings are published in each organization's listing directory or approval guide.

Factory Mutual Data Sheets are also a good reference for information and criteria for maximum foreseeable loss (MFL) fire walls. MFL walls are essentially similar to IRI's "Standard Fire Wall" (see Figure 6-9) classifications except that MFL wall designs have provisions to permit wall openings (see Tables 6-8 and 6-9). The primary difference between an MFL wall and a non-MFL wall lies in the MFL wall's ability to better maintain its structural integrity during an uncontrolled fire. Table 6-9 identifies MFL wall general design characteristics.

While other types of fire barriers and partitions discussed are generally designed to control fires in conjunction with automatic fire protection, fire departments and other measures, the MFL-type wall is designed to act alone to limit the fire.

TABLE 6-7
Fire Wall Construction Features

Wall Type	Features
Standard Fire Wall	• Freestanding concrete or masonry wall • No openings in wall. • Extends from the lowest level, through the roof, and forms a parapet 3 ft. (0.92 m) high. • Rated endwall or wingwall provided at building exterior wall where there is potential for fire to spread around the fire wall. • Reinforcement or support designed to resist damage from collapse or expansion of storage. • Wall has ability to resist fracture and penetration as a result of fire exposure. • Wall must extend from exterior wall to exterior wall or from one standard fire wall to another.
Freestanding Fire Wall	• Similar to standard fire wall except that protected openings are permitted. • No ties to a building other than metal flashing at roof level. • Masonry walls, constructed of concrete brick or poured concrete. • Sufficient clearance maintained to allow for expansion of adjacent structural components. • Masonry or other fire-rated construction encases the building's steel framing structure or is attached to the framing structure (framing then coated with fire proofing material).
Tied Fire Wall	• Roof steel on other side of wall strong enough to resist forces of roof collapse on fire exposed side. • Tied to a steel building frame on one side only. • Totally independent of the structural frame on the other side of the wall.
One Way Fire Wall	• Commonly used to subdivide a lighter hazard storage area from a higher hazard area. • Designed to withstand a fire on one side only, and tied to structural members on the low hazard side. • Clearance maintained between the one-way wall and steel on the high hazard side to prevent damage to the wall due to steel expansion. • Two one-way fire walls installed back to back. • Structural steel exposed on one side of each wall, and a specified minimum clearance provided between the two walls.
Double One-Way Fire Wall	• Used to separate an existing structure and a new addition. • Existing wall is upgraded to the desired fire resistance rating and is structurally tied to the building frame. • New addition is constructed with a second one-way wall close to the existing structure which is tied into the new buildings frame. • Only one frame and wall will collapse if an uncontrolled fire originates on either side of this double wall. • Remaining wall supported by the second structure will stay in place to stop the fire.
Source: Industrial Risk Insurers and Factory Mutual	

TABLE 6-8
Fire-Rated Separation Construction Features

Wall Type	Features
Fire Barrier	• Typically non-load bearing walls with a fire resistance rating of 2 to 3 hours. • Extend from floor to floor or floor to roof only. • Supported by either the floor and ceiling or by structural columns. • No endwalls or parapets are required. • Supporting members, such as floors, roofs, or columns, should be at least as fire resistive as the fire barrier it supports.

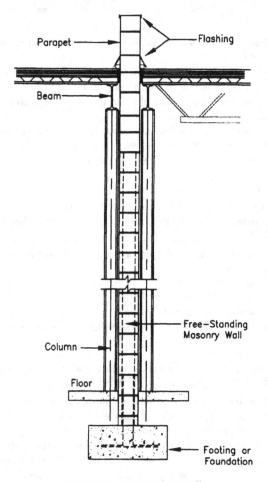

FIGURE 6-9. *MFL Fire Wall*

6.5. Fire Mitigation Construction Features

TABLE 6-9
MFL Fire Walls

Features	Characteristics
Purpose	• Relied upon to confine the fire to the area of origin until it is either extinguished or burns itself out.
Fire Resistance	• 4 hour fire resistance rating. • Only tested assemblies or materials used. • Type M or type S mortars used for masonry walls. • Protection of openings with double fire doors • Specific design criteria for through-penetrations.
Stability	• Independent of roof support/ceiling system. • Specific requirements where MFL walls are tied to building frames. • Additional reinforcing criteria. • Criteria for panel strengths.
Source: Factory Mutual	

6.5.2. Protection of Openings and Penetrations

Although fire spread can occur by heat transfer through or by structural failure of fire walls, partitions, and barriers, the most common method of fire spread is through open doors and/or unprotected penetrations in the wall itself. An unprotected opening or through-penetration can significantly affect the ability of a fire resistance rated barrier to confine a fire.

Openings of any type in fire resistive walls should be kept to a minimum. This includes openings for personnel, material handling, ducts and conveyor systems as well as through-penetrations for piping, conduit and cable. Where such openings are necessary, they should be protected by listed fire doors, shutters, dampers and through-penetration firestop assemblies. Listed fire door and shutter assemblies are tested in accordance with the criteria of ASTM E152, UL 10B, or NFPA 252.

Fire door assemblies generally consist of the fire door, the frame and the door hardware. Fire doors can be classified by an alphabetical designation (identifying the opening that is protected), an hourly designation (identifying the fire resistance rating), or a combination of the two. Building codes and NFPA 80 generally specify fire doors by their fire protection rating rather than the opening classification. Table 6-10 identifies the opening protection required for various fire resistance rated walls. Figures 6-10 and 6-11 provide examples of opening protection arrangements.

TABLE 6-10
General Opening Protection Criteria

Fire Resistance Rating of Wall	Number and Fire Resistance Rating of Door(s) and Shutters
4 Hour	Two 3-hour doors for MFL walls, fire walls and where specifically required by code, or one 3-hour door for other walls.
3 Hour	One 3-hour door, or two 1½-hour doors in a vestibule arrangement.
2 Hour	One 1½-hour door.
1 Hour	One 1-hour door for fire-rated separations enclosing shafts, or one 45-minute door for other walls.

Each opening in a fire-rated wall should be evaluated to determine that the proper fire door assembly is used. While an appropriately rated fire shutter may provide adequate opening protection, it cannot be used when the opening serves as the exit path for building occupants. Similarly, a horizontal exit in a fire-rated wall should be provided with a hinged fire door that swings in the direction of exit travel. Separate openings and rated assemblies may be required to serve personnel and material handling needs. Such an arrangement is common when protecting openings in MFL or standard fire walls.

Refer also to the "containment" section of this chapter for a discussion of liquid containment methods. Ramping may need to be provided at floor level openings in fire resistance rated barriers serving as a containment scheme component to prevent liquid travel under the fire door.

6.5.3. Through-Penetrations

Penetrations of fire resistive walls and floors are made on the basis of practical necessity or engineering design. For MFL or fire walls, these penetrations should generally be avoided.

In general, penetrations and openings in fire resistance rated walls should be fire stopped with a material having a fire resistive rating equivalent to that of the wall. A wide variety of suitable firestop materials are available. Fire stop materials are listed by recognized testing laboratories for use under specific conditions in accordance with ASTM E119 as part of a rated wall/partition assembly, or ASTM E814, "Standard Test Method for Fire Tests of Through-Penetration Fire Stops." Where permitted by the applicable building code, traditional firestopping materials should be used. For example, materials such as concrete, grout or mortar

6.5. Fire Mitigation Construction Features

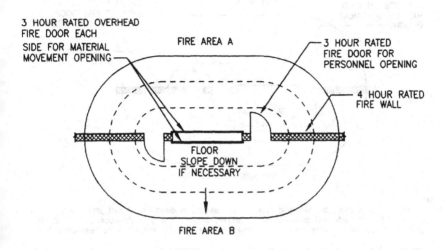

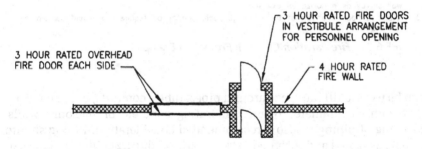

Note:
1. Personnel door(s) must be swinging type. Door swings in direction of exit travel when used as an exit. Two personnel fire doors shown, each swinging in direction of exit travel such as a condition where each door serves as a horizontal exit from its respective fire area.
2. Protection at non-personnel opening may consist of overhead fire shutter, horizontal sliding fire door or other rated fire door assemblies that are listed for opening protection.
3. Where the authority having jurisdiction or insurance carrier requires rated door assemblies on each side of the fire wall, personnel doors may be installed in a vestibule arrangement.
4. A sloped floor is shown at the openings such as may be required for liquid containment purposes.

FIGURE 6-10. *Fire Wall Opening Protection Examples*

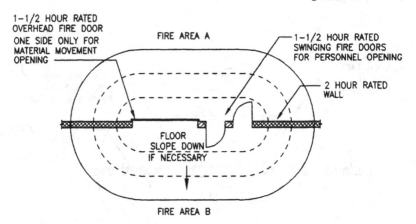

Note:
1. Personnel doors must be swinging type. Door swings in direction of exit travel when used as an exit. Opposite swinging fire doors shown, each swinging in direction of exit travel such as a condition where each door serves as a horizontal exit from its respective fire area.
2. Overhead fire door must be arranged for automatic closing when the fire resistance rated wall serves as a horizontal exit wall.
3. A sloped floor is shown at the openings such as may be required for liquid containment purposes.

FIGURE 6-11. *Fire Partition Opening Protection Example*

can be used to fill the space around pipes, tubes, conduits, etc., up to 6 in. (150 mm) in diameter when penetrating concrete or masonry walls. Sleeving of piping is also a recommended consideration. Piping should also be designed and installed to "break-away" during ceiling collapse, or be located less than 3–4 feet above the finished floor in MFL walls.

The choices of fire stop products include:

- Intumescent products suitable even for plastic penetrating items.
- Sealants that are either solvent-based or water-based formulations.
- Combinations of materials and components such as silicone foams, fire resistive board materials, mortars, collars and wire mesh.
- Factory-built fire stop devices intended to be installed around a penetrating item, such as a collar assembly that is attached to the wall or floor surface.

A fire stop listing or approval will describe the various details related to the construction and assembly of the fire stop which may include:

- Floor or wall construction and thickness.
- Fire performance rating (F-rating).

6.5. Fire Mitigation Construction Features

- Heat transmission performance rating (T-rating).
- Air leakage rating (L-rating).
- Type and size of penetrating item.
- Maximum area of opening.
- Allowable annular space dimension.
- Materials and devices required to fill the opening or annular space.

When selecting an ASTM E814 tested fire stop, the details indicated by the listing or approval are technically required to be provided as indicated. However, details of a tested system frequently do not match actual design parameters or field installed conditions. Reasonable care should be exercised when using a fire stop design to ensure that the field application is similar in opening size, annular space size, wall or floor construction, and types and sizes of penetrations. The fire stop product manufacturer should be contacted for assistance and/or engineering judgments regarding the expected performance of modified fire stop details.

The designer will need to consider whether the fire stop will:

- Serve as a water seal.
- Serve as a smoke seal.
- Impair the ability for penetrating items to be installed or removed periodically.
- Be capable of accommodating future modifications.
- Need to be painted.
- Have a curing time that meets schedule constraints.
- Be compatible with the environmental conditions such as standing water, high humidity or chemical atmospheres.
- Be subject to movement or vibrations.

6.5.4. Heat and Smoke Venting

Hot smoke and gases generated by a fire will rise within a warehouse until their upward movement is limited by the roof. They will then flow outward in a radial pattern along the underside of the roof unless walls or other building members are encountered. As this process continues, hot smoke and gases will begin descending toward the floor and eventually fill the interior of the warehouse. Due to the resultant loss of visibility and higher temperatures, rescue and manual firefighting operations may be impeded.

For unsprinklered warehouses, NFPA 204M, "Guide for Smoke and Heat Venting," has criteria for the installation of draft curtains and automatic smoke and heat vents. Draft curtains can restrict the horizontal flow of hot gases and smoke beneath a roof. When used in conjunction with automatic heat and smoke vents, they can facilitate the discharge of hot gases and smoke.

Roof mounted heat and smoke vents can be categorized as either gravity or mechanically operated. The design criteria in NFPA 204M is based upon the predicted heat-release rate for a number of commodities.

The use of draft curtains and automatic heat and smoke vents in sprinklered warehouses is not addressed in NFPA 204M. Fire tests involving various stored commodities using operational sprinkler systems and automatic heat and smoke vents has been limited and inconclusive.

Fire tests comparing sprinklered occupancies with and without draft curtains revealed that draft curtains had a negative impact on the sprinkler system performance. Depending on the location of the fire relative to the ceiling sprinklers and draft curtains, draft curtains may interfere with the spray pattern of the sprinklers. If the conical spray pattern of the sprinklers impinges on an adjacent draft curtain, it will interfere with the control or suppression capabilities of these sprinklers. This may result in a larger fire size with a resultant higher number of operating sprinklers and greater water usage. In addition, excessive numbers of automatic sprinklers may operate, overtaxing the water supply capability, increasing the run-off requiring containment and treatment, and increasing product damage, while providing little additional protection.

Sprinkler system design criteria in standards published by the NFPA and FM were developed without the use of draft curtains and automatic heat and smoke vents. FM does not require heat and smoke vent installation. However, FM Loss Prevention Data Sheet 8-9, "Storage of Class 1, 2, 3, 4, and Plastic Commodities," recommends that heat and smoke vents be manually actuated in sprinklered warehouses, if installed. Conversely, IRI recommends automatic heat and smoke vents in sprinklered buildings except where early-suppression fast-response (ESFR) sprinkler systems are used.

If automatic operation is required by local codes, then the operating temperature of the sprinklers should be lower than that of the heat and smoke vents. This is particularly relevant when fast-response, ordinary temperature rated sprinklers are installed. FM Loss Prevention Data Sheet 8-9 also recommends, if required by local codes, that draft curtains be placed midway between sprinklers and sprinkler branch lines. If this is not feasible, then additional sprinklers may be required.

6.6. Deflagration Prevention and Mitigation

6.5.5. Powered Ventilation Systems

Powered ventilation systems in chemical warehouses are typically used for some flammable gas and liquid storage areas (see Gas and Vapor Control, Section 6.6). Heating and ventilation systems and powered roof exhaust systems can be arranged to shut down automatically upon receipt of a fire alarm signal so as not to interfere with sprinkler system operations due to airflow. Furthermore, these systems can also be arranged for manual starting and full exhaust by the fire department for smoke removal during search and rescue operations.

Some stored commodities, such as toxic liquids, may warrant a special ventilation system and compartmentation of storage to assure that toxic vapors that may be released do not accumulate in an area.

6.5.6. Emergency and Standby Power Systems

Where emergency and standby power systems are included in the warehouse design, NFPA 110, "Emergency and Standby Power Systems," provides criteria for the design and installation of such systems. Such systems should be considered to back up critical functions at sites where the electrical power supply is unreliable. Applications for this back up power can include: emergency lighting (See Chapters 4 and 6), communication systems, refrigeration systems, mechanical ventilation systems, and vapor/gas detection systems (See Chapter 6), and electric fire alarms and fire pumps (See Chapter 7).

6.6. Deflagration Prevention and Mitigation

Certain materials can deflagrate or explode under adverse conditions. These include several organic peroxide formulations and oxidizers, compressed or liquefied flammable gases and flammable liquid vapors.

Class I organic peroxides (NFPA 43B) can decompose violently or explosively when heated. Low temperatures can cause some organic peroxide formulations to separate into different phases. Organic peroxide crystals and the carrier can then produce a mixture that is more unstable than the original formulation. Most organic peroxide formulations can generate flammable gases, vapors or mists during decomposition and generate an explosive atmosphere within an enclosure.

When subjected to high temperature, some Class IV oxidizers (NFPA 430) can decompose and deflagrate. Additionally, compressed or liquefied flammable gases and flammable liquid vapors, especially Class IA liquids, can also produce an explosive atmosphere if released within a building or room. Chapter 2 provides additional details regarding the deflagration potential of various commodities.

Measures can be taken to prevent or mitigate the effects of a deflagration. Decomposition reactions and phase separations can be prevented by control of the storage temperature. Under normal conditions this may require the use of a refrigeration or heating system. In the event of a fire, high temperatures may be limited with some materials through the use of a sprinkler system. The risk of an explosive atmosphere involving flammable gases or vapors can be reduced through packaging design and/or by mechanical ventilation. Finally, ignition of explosive atmospheres can be prevented by eliminating all ignition sources. As people can introduce unexpected ignition sources, such a strategy may prove challenging.

Deflagration hazards can be mitigated by storing sensitive materials in adequately separated detached buildings. As an alternate to spatial separation, the warehouse and attached or detached buildings/enclosures can incorporate damage limiting construction into the design.

6.6.1. Temperature Control

Chemicals should be stored within the temperature range recommended by the manufacturer. Proper temperature control may be important for product quality and can also have an impact on the safe storage of these materials. While this consideration should apply in general, some classes of chemicals have special needs.

A refrigeration system may be necessary for storage areas containing materials subject to a heat-induced decomposition reaction. These storage areas should be maintained at a temperature well below the self-accelerating decomposition temperature (SADT). Also, materials subject to phase separation resulting in more unstable mixtures should be maintained within the recommended temperature range. The recommended temperature range for a given commodity can be obtained from the manufacturer or the material safety data sheet (MSDS).

Localized heating or cooling is also undesirable and should be avoided. Adequate space separation should be maintained between heat sensitive chemicals and heating and cooling coils, pipes, ducts and diffusers. Refrigeration systems that utilize nonflammable gases or indirect expansion of flammable gases and heating systems that use hot water, low pressure steam, or indirectly heated warm air are preferable.

6.6. Deflagration Prevention and Mitigation

In addition to the normal temperature control provided by refrigeration and heating systems, separate temperature alarms should be considered. Both low and high temperature limit switches can be connected into the facility alarm system providing a more reliable means of monitoring area temperature. Finally, emergency generators should be considered where electrical power is critical for temperature control and the primary source is not reliable.

Chemicals sensitive to high temperatures from fire exposure can be cooled by the wetting effects of a sprinkler system. This would include flammable liquid monomers, such as acrylonitrile, and Class IV oxidizers. A properly designed deluge sprinkler system is an effective means for cooling fire exposed containers. These systems are discussed in greater detail in Chapter 7. Materials having a decomposition temperature that is lower than the sprinkler water temperature (e.g. certain refrigerated organic peroxides) would not be good candidates for sprinkler protection, as the discharge of water onto this material may initiate or accelerate a decomposition reaction. Gaseous or dry chemical agents may be better suited, depending on their compatibility, than water.

Materials with low thermal decomposition onset temperatures may cause rapid expansion of the "fire area" and may overtax a sprinkler system. Spatial separation in the form of aisles between these materials can become critical.

6.6.2. Gas and Vapor Control

An explosive atmosphere will exist within an enclosure when there is sufficient flammable gas or vapor present to raise the concentration of vapor above the lower flammable limit (LFL). The most severe deflagration potential exists when these flammable gases and vapors fill the enclosure volume slightly above the stoichiometric concentration (C_{st}). It should be emphasized, however, that the entire enclosure volume does not have to be within the flammability limits for an explosive atmosphere to exist. A deflagration can be produced when only a fraction of the total room volume is within the flammable range. In a storage occupancy, compressed or liquified gases or flammable liquids must be released from their cylinders or containers through some accidental means. Furthermore, flammable liquids must also evaporate and produce vapor at a sufficient rate within the enclosure. Under fire conditions containers may become pressurized sufficiently to violently rupture, rapidly releasing a flammable vapor cloud. In a storage occupancy a release from a cylinder or container could result from physical damage or pressure buildup.

Operations such as filling, dispensing and sampling should not be performed in the warehouse area. Provisions should be made for handling damaged or leaking containers. Section 9.6 addresses emergency spill response.

In general, containers of pressurized flammable gases should be stored in outdoor open air areas with overhead protection from the elements. If outside storage is not an option, controlling the buildup of flammable gases or vapor can be accomplished, up to a limit, with either a passive or active ventilation system. However, a large release of gas or vapor may exceed the capacity of a practical ventilation system resulting in a hazardous concentration.

The passive systems, known as gravity or natural draft type, rely on convection and the wind for air movement. The active systems are mechanical and generally provide a more reliable means for ventilation. The layout of the system should depend upon the physical properties of the gases or liquids stored. Gases that are lighter than air will rise within the enclosure and should be ventilated at the top of the enclosure. Conversely, heavier than air gases and vapors will require low level ventilation for removal.

Natural ventilation can be adequate provided there are sufficient vents or louvers at the ceiling or floor level. An effective design will ensure that the airflow either sweeps across the floor or ceiling continuously and adequate outside make-up air is provided. The discharge should be directed outside and away from any air inlets, other openings, and equipment such as compressors. If adequate natural ventilation cannot be provided a mechanical ventilation system should be used. Mechanical ventilation systems for applications involving transfer of flammable liquids typically have a capacity of 0.25 to 2 cfm/ft^2 (0.075–0.60 m^3/min/m^2) and higher in trenches, sumps, or other collection areas.

Where deemed necessary, flammable gas detection can be installed as an early warning mechanism.

6.6.3. Sources of Ignition

If an explosive atmosphere exists, ignition can be prevented by elimination and control of all potential ignition sources. Open flames, improper electrical equipment and industrial trucks should not be allowed in areas where there is potential for an explosive atmosphere. In some cases, autoignition can also occur during decomposition or contact with temperature extremes. Chapter 3 contains additional information on control of ignition sources.

6.6. Deflagration Prevention and Mitigation

6.6.4. Spatial Separation

Adequate spatial separation can be an effective and economical means to mitigate the effects of conventional fire propagation or a deflagration. If space is available, the thermal energy, pressure wave and any projectiles produced during a deflagration can be dispersed, thereby reducing exposure to people and important buildings. (NFPA 430 and 43B have spatial separation recommendations for detached buildings containing Class IV oxidizers and Class I organic peroxide formulations respectively.)

6.6.5. Damage Limiting Construction

If adequate spatial separation cannot be provided, severe damage to adjoining or nearby portions of a warehouse due to a deflagration can be prevented through damage limiting construction. This approach to deflagration mitigation includes the use of:

- no containment (e.g., no walls)
- all walls pressure-relieving
- all walls pressure-resistant
- a combination of pressure-relieving and pressure-resistant walls.

When properly designed and constructed, pressure-resistant walls and doors will withstand deflagration forces. Also, pressure-relieving walls discharge gases and maintain internal pressure below the damage threshold. Damage limiting construction will not prevent or minimize injury to personnel located within the enclosure. Also, as the name implies, these features can be effective against deflagration, but not detonation-type explosions.

The design of damage limiting construction is not an exact science. There are two widely accepted methodologies for the design of damage limiting. These methodologies are described in NFPA 68, "Standard for Venting of Deflagrations" and FM Loss Prevention Data Sheet 1-44, "Damage-Limiting Construction." These standards are directly applicable to flammable gases and flammable and combustible liquids. Additional specific criteria for organic peroxide formulations can found in NFPA 43B.

The need for damage limiting construction is not universally accepted for all materials having a deflagration potential especially where storage of hazardous materials is involved. For example, the deflagration potential associated with closed container storage of Class IA liquids is not considered sufficiently high by some insurance companies to warrant this protection. NFPA considers Class IA liquids in closed 1 gallon (3.8 l) and larger containers to present a deflagration hazard under adverse conditions.

A building or enclosure incorporating pressure resistant and pressure relieving features should be designed by an engineer competent in this area. Construction should be performed by qualified construction contractors. Vent panels and vent panel relief devices that have been listed or approved for this use are available.

Figure 6-12 depicts various building layouts incorporating spatial separation and damage limiting construction.

6.7. Natural Peril Mitigation

If a warehouse site is prone to exposure from one or more natural perils, design features may be incorporated which mitigate the hazards. Damage potential from earthquake, flood, lightning or windstorm/hurricane may thus be reduced or eliminated. The design and construction features discussed in this section can be applied to new, and in many cases, existing facilities.

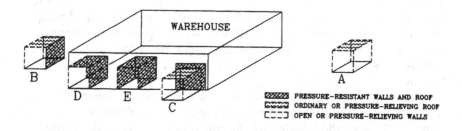

Locations for areas having a deflagration potential in order of preference:
    A: Detached building having adequate spatial separation from warehouse. No walls or pressure-relieving walls on all sides. Ordinary or pressure-relieving roof.
    B: Detached building without adequate spatial separation from warehouse. No walls or pressure-relieving walls on three sides and pressure-resistant wall facing warehouse. Ordinary or pressure-relieving roof.
    C: Attached building with no walls or pressure-relieving walls on three sides and pressure-resistant wall facing warehouse. Ordinary or pressure-relieving roof.
    D and E: Enclosure within warehouse on exterior wall(s). Exterior wall(s) pressure-relieving and interior wall(s) and roof pressure resistant.

FIGURE 6-12. *Deflagration Mitigation Using Adequate Spatial Separation or Damage-Limiting Construction and Deflagration Venting*

6.7.1. Earthquake

Seismic designs incorporate features that dissipate or absorb the energy released during an earthquake in a controlled fashion. Despite the use of seismic design features, however, an earthquake-proof structure cannot be guaranteed due to the unpredictable nature of earthquakes. All three of the model building codes used in the U.S contain provisions for designing seismic structures. The provisions in all three codes are similar; however, the Uniform Building Code is currently the most widely used in the U.S. for seismic design loading. The structural design parameters in this code are determined considering the seismic zone, site characteristics including soil type profile, occupancy characteristics, building configuration requirements, and classification of structural system. Table 6-11 contains examples of earthquake mitigation features

TABLE 6-11
Earthquake Mitigation Features

Warehouse Building	• Provide seismic design of warehouse structural framing system using one of four types: 　• Bearing wall system 　• Building frame system 　• Movement-resisting frame system 　• Dual system
Storage Racks	• Anchor rack columns to concrete floor with larger baseplates. • Reinforce column-to-beam connectors. • Double the number of columns. • Post weight limits on racks and do not exceed maximum permissible weights. • Avoid top heavy racks and double stacking upper pallets. • Store most hazards materials on lower levels.
Fire Sprinkler System	• Use flexible couplings and swing joints on piping. • Provide adequate clearance around pipes penetrating walls, floors, and foundation. • Provide lateral and longitudinal sway bracing. • Use diesel drivers for fire pumps. • Use of PVC underground pipe.
Building Utilities	• Provide fuel gas supply lines with seismic-actuated shutoff valves. • Locate building unit heaters and associated piping external to building. • Provide remote shutoff for electrical service.

for the warehouse structure, storage racks, fire sprinkler system, and building utilities.

6.7.2. Flood

Flood mitigation features can be categorized as permanent, contingent or emergency protection. Permanent features are integral components of the warehouse design or site layout and require no human intervention to be effective. Contingent protection is less reliable as it requires human intervention and sufficient warning time. The least effective response to a flood exposure is emergency protection which can require considerable human intervention and warning time.

Design considerations should include hydrostatically unbalanced forces associated with floodwater pressure applied to one side of a flood barrier. The barriers that can be subject to these forces include both floors and walls. Additionally, hydrodynamic forces associated with flowing water should also be considered. Table 6-12 provides examples of flood mitigation features that may be used for permanent, contingent, and emergency protection.

TABLE 6-12
Flood Mitigation Features

Permanent Protection	• Design lowest floor of warehouse and access roadway above base flood level. • Build earthen levee or concrete flood wall with sufficient free board.
Contingent Protection	• Provide watertight closures • Flood doors. • Flood shields over doors and windows. • Provide manually operated backflow valves on floor drains and sewer lines.
Emergency Protection	• Remove stored commodity requiring special handling, particularly hydroscopic, or water reactive as well as material requiring refrigeration. • Relocate stored commodity not requiring special handling to higher elevation such as upper floor or upper tier of steel storage rack. • Sandbag site or warehouse openings.

6.7.3. Lightning

Depending on the risk of loss due to lightning, a lightning protection system may be warranted to protect a warehouse from damage. A method for assessing the risk is presented in NFPA 780, "Lightning Protection Systems." This standard also outlines various design alternatives for lightning protection of buildings and includes lightning conductors, air terminals, ground terminals and associated fittings. Absent a lightning protection system, the building itself may require grounding of building steel.

6.7.4. Windstorm, Hurricane and Tornado

Incorporating wind resistant construction features into the design of a warehouse can reduce the damaging effects of severe wind forces. Although wind damage can be incurred by any part of the building, the loss history has shown that most losses involve roof failures. These failures are usually the result of an improperly designed or constructed roof. A method for estimating design wind forces on buildings can be found in FM Loss Prevention Data Sheet 1-7, "Wind Forces on Buildings and Other Structures."

Design based upon the FM methodology can be effective against windstorms having up to hurricane force winds. They are also effective against wind forces generated by tornadoes categorized as F-2 and lower on the Fujita Tornado Wind Intensity Scale.

Table 6-13 contains examples of windstorm mitigation features including those that pertain to roofs.

6.8. Security Features

Preventing unauthorized access into a warehouse may require certain security features. Unauthorized access may be motivated by theft or other illegal or destructive purposes. There may also be a need to restrict employee access into an area of a warehouse storing toxic or potentially hazardous materials. The type and extent of security features utilized should be based upon the risk associated with unauthorized entry. Security features, such as door locks, alarm systems, closed circuit television (CCTV) monitors, should not interfere with required means of egress.

Table 6-14 offers a list of passive and active security features.

TABLE 6-13
Windstorm Mitigation Features

Permanent Protection		
	New Warehouses	• Reduce exposure by selecting a site that is not surrounded by or adjacent to large, open flat areas or hurricane/high wind prone areas.
		• Use structural concrete roof decks in hurricane prone areas.
		• Design and install insulated steel deck roofs in accordance with the UL Roofing Materials and Systems Directory or the FM Approval Guide and appropriate FM data sheets.
		• Avoid large expanses of glass; use blank walls or narrow profile windows.
		• Use continuous hook strip below and parallel to flashing and firmly secure to building.
	Existing Warehouses	• Reinforce insulated steel deck roofs with additional mechanical roof fasteners.
		• Use continuous hook strip below and parallel to flashing and firmly secure to building.
		• Replace glass with break-resistant glazing materials such as polycarbonate, acrylic, or laminated glass.
Contingent Protection		• Cover windows with hurricane shutters.
		• Minimize water damage potential by skidding storage off floor.

6.9. Outdoor Storage

Outdoor storage will not be feasible for many commodities due to quality control or risk management issues. These commodities may require the controlled environment and building protection features only possible within an enclosed warehouse.

However, outdoor storage may be desirable for other commodities having certain properties. For example, a leaking toxic gas cylinder, in an enclosed warehouse, would present an unacceptable and unmanageable risk.

If outdoor storage is planned, then consideration should be given to provision of certain features. These features, which are outlined in this guideline, may be warranted for exposures such as environmental, fire, or security. Table 6-15 offers NFPA references for "commodity specific" outdoor storage recommendations.

TABLE 6-14
Security Features (Options)

Passive Protection	• Exterior lighting. • Exterior fencing. • Buffer zones • Window bars. • Burglar-resistant skylights. • Door locks.
Active Protection	• Guard services. • Alarm protection annunciated locally and/or at a central station. • Contacts on doors and windows. • Glass break detectors. • Photoelectric beams. • Card access system. • Motion detectors. • Infrared beams. • Sound detectors. • CCTV monitoring.

TABLE 6-15
Outdoor Storage References

Commodity Type	NFPA Code or Standard
Aerosol Products	30B
Compressed and Liquefied Gases in Portable Cylinders	55
Ethylene Oxide	560
Flammable and Combustible Liquids	30
Liquid and Solid Oxidizers	430
Ordinary Commodities, Plastics, Elastomers and Rubber	231

References

American Society for Testing and Materials, "Fire Tests of Door Assemblies." ASTM E152, Philadelphia, PA, 1990.
American Society for Testing and Materials, "Standard Methods of Fire Tests of Building Construction and Materials." ASTM E119, Philadelphia, PA, 1988.

American Society for Testing and Materials, "Standard Test Method For Fire Tests of Through-Penetration Fire Stops." ASTM E814, Philadelphia, PA, 1983.
American Society for Testing and Materials, "Standard Test Method for Determining Effects of Large Hydrocarbon Pool Fires on Structural Members and Assemblies." ASTM E1529, Philadelphia, PA, 1993.
Building Officials & Code Administrators International, Inc., "The BOCA National Building Code." Country Club Hills, IL, 1993.
Building Officials & Code Administrators International, Inc., "The BOCA National Fire Prevention Code." Country Club Hills, IL, 1993.
Building Officials & Code Administrators International, Inc., "The BOCA National Mechanical Code." Country Club Hills, IL, 1993.
Building Officials & Code Administrators International, Inc., "The BOCA National Plumbing Code." Country Club Hills, IL, 1993.
Factory Mutual Engineering Corporation, "Damage-Limiting Construction." Loss Prevention Data Sheet 1-44, Norwood, MA, July, 1991.
Factory Mutual Engineering Corporation, "Drainage Systems for Flammable Liquids." Loss Prevention Data Sheet 7-83, Norwood, MA, March, 1991.
Factory Mutual Engineering Corporation, "Storage of Class 1, 2, 3, 4 and Plastic Commodities." Loss Prevention Data Sheet 8-9, Norwood, MA, September 1993.
Factory Mutual Engineering Corporation, "Wind Forces on Buildings and Other Structures." Loss Prevention Data Sheet 1-7, Norwood, MA, April 1983.
Industrial Risk Insurers, "Fire Walls, Fire Barriers and Fire Partitions." IRInformation IM.2.2.1, June 1, 1992.
International Conference of Building Officials, "Uniform Building Code." Whittier, CA, 1994.
International Conference of Building Officials, "Uniform Fire Code." Whittier, CA, 1994
International Conference of Building Officials, "Uniform Mechanical Code." Whittier, CA, 1994.
International Conference of Building Officials, "Uniform Plumbing Code." Whittier, CA, 1994.
Midwest Plan Service, "Designing Facilities for Pesticide and Fertilizer
Containment." Ames, IA, 1991.
National Fire Protection Association, "Flammable and Combustible Liquids Code." NFPA 30, Quincy, MA, 1996.
National Fire Protection Association, "Code for the Storage of Organic Peroxide Formulations." NFPA 43B, Quincy, MA, 1993.
National Fire Protection Association, "Guide for the Venting of Deflagrations." NFPA 68, Quincy, MA, 1994.
National Fire Protection Association, "National Electric Code." NFPA 70, Quincy, MA, 1996
National Fire Protection Association, "Standard for Fire Doors and Fire Windows." NFPA 80, Quincy, MA, 1995.
National Fire Protection Association, "Life Safety Code." NFPA 101, Quincy, MA, 1997
National Fire Protection Association, "Standard for Emergency and Standby Power Systems." NFPA 110, Quincy, MA, 1996.
National Fire Protection Association, "Guide for Smoke and Heat Venting.", NFPA 204M, Quincy, MA, 1991.
National Fire Protection Association, "Standard on Types of Building Construction." NFPA 220, Quincy, MA, 1995.
National Fire Protection Association, "Standard Methods of Fire Tests of Door Assemblies." NFPA 252, Quincy, MA, 1995.

Additional Reading

National Fire Protection Association, "Code for the Storage of Liquid and Solid Oxidizers." NFPA 430, Quincy, MA, 1995.
National Fire Protection Association, "Standard for the Storage, Handling, and Use of Ethylene Oxide for Sterilization and Fumigation." NFPA 560, Quincy, MA, 1995.
National Fire Protection Association, "Lightning Protection Systems." NFPA 780, Quincy, MA, 1993.
Nguyen, Duy T. and Michael F. Broder, Cathy L. McDonald and Eugene Zarate, "Protective Secondary Containment from Pesticides and Fertilizer." *Journal of Protective Coatings and Linings*, July 1995.
Southern Building Code Congress International, Inc., "Standard Building Code." Birmingham, AL, 1994.
Southern Building Code Congress International, Inc., "Standard Fire Prevention Code." Birmingham, AL, 1991.
Southern Building Code Congress International, Inc., "Standard Mechanical Code." Birmingham, AL, 1991.
Southern Building Code Congress International, Inc., "Standard Plumbing Code." Birmingham, AL, 1991.
Southern Building Code Congress International, Inc., "Standard Gas Code." Birmingham, AL, 1991.
Tabar, David C., "Construction and Containment Design Considerations for Hazardous Materials Warehousing." *Spray Technology and Marketing*, September, 1996
Tabar, David C., "Environmental and Construction Design Considerations for Hazardous Material Warehousing," AIChE Loss Prevention Symposium, Boston, MA, August 1, 1995.
Underwriters Laboratories Inc., "Fire Tests of Door Assemblies." UL 10B, Standards for Safety, September 28, 1992.
Underwriters Laboratories Inc., "Rapid Rise Fire Tests of Protection Materials for Structural Steel." UL 1709, Standards for Safety, February 27, 1991.

Additional Reading

Aldinger, T.I., "Designing Secondary Containment and Other Concrete Structures to Optimize Coatings Performance." *Journal of Protective Coatings and Linings*, August 1992.
Aldinger, T.I. and B.S. Fultz, "Selecting Coatings and Linings for Concrete in Chemical Environments." *Journal of Protective Coatings and Linings*, August 1992.
AIA Research Corporation, "Design Guidelines for Flood Damage Reduction." Federal Emergency Management Association FEMA-15, Washington, DC, October 1981.
"Concrete Construction Annual Reference Guide." Concrete Construction Publications, Inc., Addison, IL.
"Cracks in Concrete: Causes and Prevention." Concrete Construction Publications, Inc., Addison, IL.
Federal Emergency Management Agency, "Floodproofing Non-Residential Structures." FEMA-102, May 1986.
Ilaria, Joseph, "Prolong the Life of Concrete." *Chemical Engineering Progress*, July 1995
Lathrop, James K., "Life Safety Code Key to Industrial Fire Safety." *NFPA Journal*, 88(4), 36, July/August 1994.
McGovern, G.J. and P.T. Vatala, "Storing Hazardous Materials." *NFPA Journal*, 88(5), 71, September/October 1994.

7
Fire Protection Systems

7.1. Synopsis

Fire in a chemical warehouse can have serious consequences affecting people, the environment, property, and business continuity. An effective strategy for fire mitigation requires that these consequences be understood. Chemical warehouse designers and operators can select from two basic fire mitigation strategies. This selection depends on management's philosophy and the requirements of the authority having jurisdiction. Under certain circumstances, it may be acceptable to provide no fixed fire protection systems and rely on compartmentation by fire walls or spatial separation to limit the potential growth of a fire. The other strategy is to provide fixed fire protection.

Fire protection systems may mitigate the effects of fire either by control, suppression, or extinguishment. For this strategy to be effective, several issues relative to fire severity must be considered during the design stage of the warehouse and throughout its operation, including the fire hazard of the commodity and storage configuration. Fire alarm systems can provide early warning of a fire and assist in evacuation of the warehouse. Additionally, emergency responders and warehouse management can also be alerted. This can be of particular importance if a strategy of fire control is employed. Finally, the potential for fires can be minimized by controlling sources of ignition.

Manual fire suppression is not normally dependable as a primary fire protection strategy for chemical storage warehouses. Fires may grow to uncontrollable size before effective manual response can be employed, and may pose severe risks to firefighters. Moreover, manual suppression usually involves far more water application than automatic systems, aggravating problems of disposal of fire water runoff.

7.2. Storage Considerations

Fire growth throughout a storage array is a function of a number of conditions, including the intrinsic fire hazard of the commodity, storage height and storage configuration. The intrinsic fire hazard of commodities can be ranked by various systems of commodity classification as discussed in Chapter 2.

Higher storage heights produce more severe fires than lower storage heights of the same commodity. For protected warehouses, sprinkler system design criteria is specific to a range of storage heights. Sprinkler system design criteria will be more demanding as storage height increases. Also, greater clearance between the top of storage and ceiling sprinklers will result in more severe fires than with smaller clearances due to slower operation of ceiling sprinklers and greater travel distances for sprinkler water discharge through hot combustion gases.

Open frame steel rack storage arrays allow for greater exposed surface area of the commodity and greater pile stability during a fire than solid pile and palletized storage arrays. For these reasons, rack storage arrays will produce more severe fires than other types of storage arrays under similar conditions. This phenomena is most noticeable in multi-row rack arrays where horizontal fire spread is most severe. Single and double-row racks have aisle spaces around them that can provide a fire break under certain conditions.

Compartmentation within a rack array, consisting of horizontal and vertical barriers, can be used in conjunction with in-rack sprinklers in each rack bay. This combination can result in smaller and more easily controlled, suppressed, or extinguished fires with fewer operating sprinklers. These barriers will limit fire travel down the length of a rack as well as upward fire growth allowing for faster operation of in-rack sprinklers. This can have the added benefits of reduced quantity of contaminated sprinkler water runoff and reduced product damage.

Automated warehouses can present unique conditions not found in warehouses where material handling is done manually. Typically, these facilities use rack storage heights that are much higher and present a greater fire challenge requiring special fire protection considerations. Constant attendance may be missing due to the emphasis on automation which may limit early detection. Furthermore, an automatic rack retrieval system has the potential to inadvertently move burning or leaking commodity to other parts of a warehouse. Consideration should be given to automatic shut down of this equipment in the event of fire.

7.3. Fire Control, Suppression, and Extinguishing Systems

A fire control, suppression, or extinguishing system can enhance life safety, minimize fire damage to a warehouse and its contents and prevent chemical releases that could potentially expose people and the environment. Chemical releases include the actual material stored, contaminated fire water run-off, and the products of combustion, decomposition or adverse reactions.

A fire is considered controlled if the protection system limits its growth to a relatively small area. A suppressed fire is one where the protection system limits fire growth to an even smaller area, resulting in a very high level of control. A controlled or suppressed fire will become fully extinguished only after manual fire fighting efforts are successful or when the fire had burned itself out in the area of fire origin. A fire is considered extinguished when the combustion process has been terminated.

In general, if the system is designed to provide control or suppression with only a few sprinklers operating, there will be less contaminated fire water run-off to be collected and treated, and less commodity damage. However, some commodity configurations require a greater number of sprinklers to achieve control or suppression.

The selection of an approach to fire protection - to provide fire protection systems for fire control, suppression, or extinguishment, or to forgo fixed fire protection systems—is influenced by several factors. The authority having jurisdiction, potential community or environmental exposures, investment at risk, insurance considerations and business continuity must be considered in reaching a decision. Without dependable fixed fire protection, fire mitigation will rely upon compartmentation and spatial separation to limit the magnitude of the fire. The selection and design of fire walls and spatial separation considerations are discussed in Chapter 6. The principal fire protection agents and systems used to achieve fire control or suppression in chemical warehouses are listed in Table 7-1.

The rate of fire growth and the ability of a fire protection system to control, suppress, or extinguish a fire depends on a number of variables. These include the ignition source, physical and combustion properties of the commodity, type of packaging system, storage arrangement, warehouse arrangement and construction features, and fire protection system design parameters.

Several parameters must be considered when selecting a fire protection agent and system. These include their effectiveness, limitations, advantages, reliability, costs and whether the goal is fire control, suppression, or extinguishment.

TABLE 7-1
Fire Protection Agents and Systems Commonly Found in Chemical Warehouses

Agent	Systems
Water	• Automatic closed-head sprinkler • Wet • Dry • Preaction • Automatic deluge sprinklers.
Low Expansion Foam Protein and fluoroprotein foam Aqueous film-forming foam (AFFF) Film-forming fluoroprotein (FFFP)	• Automatic deluge sprinkler • Automatic closed-head sprinklers • Automatic deluge sprinklers
Medium and high expansion foam	• Total flooding
Gaseous agents	• Total flooding
Dry chemical	• Total flooding
Combustible Metal Agents	• Manually applied

7.3.1. Fire Control

Fire control can be achieved for a number of hazardous chemicals by an automatic sprinkler system using "spray sprinklers" operating directly over the origin of the fire and the surrounding area. Water that is discharged from sprinklers extinguishes burning materials and pre-wets the surrounding commodity to prevent or slow the spread of fire. This is most effective with material that is water absorbent such as cartoned commodities. For other commodities, such as flammable and combustible liquids in containers or drums, sprinklers protect by cooling to prevent overpressure and container rupture.

The convective updraft from severe fires may prevent water droplets from reaching the seat of the fire and commodity may continue burning until it has been consumed or manually extinguished. Sprinklers operating directly over the fire and surrounding area will lower the ceiling temperature, thus limiting damage to the warehouse structure. Fires

7.3. Fire Control, Suppression, and Extinguishing Systems

involving certain hazardous chemicals, however, may not be controlled using water-based sprinkler systems.

7.3.2. Fire Suppression

Some commodities in certain storage configurations can be suppressed if early-suppression fast-response (ESFR) sprinklers are utilized. ESFR sprinklers will discharge a greater volume of water using larger water droplets at a higher velocity. These fast response sprinklers are also designed to operate sooner than ordinary sprinklers while the heat release rate of the fire is still relatively small. The net effect is that fewer sprinklers operate and water will penetrate down into the seat of the fire resulting in suppression. Table 7-2 offers some selection parameters associated with water as a fire protection agent.

Quick response in-rack sprinklers in addition to ESFR or large drop sprinklers at the ceiling level can greatly reduce the total number of sprinklers required to attain control or suppression. In extreme cases, compartmentalization of the racks and in-rack sprinklers for each compartment can stop the spread of fire to adjacent racks or compartments.

Palletized or solid pile storage arrays can be protected with ceiling mounted sprinkler systems. Depending on storage conditions and system design, rack storage arrays can also be protected with ceiling mounted or combination ceiling mounted and in-rack sprinkler systems. Table 7-3 offers NFPA references for "commodity specific" sprinkler design criteria including water and low expansion foam.

Water supplies for sprinkler systems can be provided by public or private systems, or a combination of both. In both instances, it is customary to also provide for a fire hose flow allowance. This allowance is particularly important with fire control sprinkler systems. Public systems can be augmented with fire pumps typically powered by a diesel or electric driver. Private systems are usually supplied by automatic fire pump(s) taking suction from a tank, reservoir, or natural body of water such as a river. The water supply should be capable of meeting the hydraulic requirements of the sprinkler system, have an adequate duration, and be considered reliable. In some cases, it may also be necessary to have two sources of water to improve the reliability of the system. This is particularly relevant in areas having an unreliable public water supply. Fire pumps should be automatic starting. If a reliable electrical supply is not available, a diesel driven fire pump should be used.

Floor drains can limit the spread of burning flammable and combustible liquids on the warehouse floor reducing the overall fire size. With-

TABLE 7-2
Water[1]

Protection Mechanism	Recommended for Fires Involving	Not Recommended for Fires Involving	Limitations	Advantages
• Cooling • Pre-wetting surrounding commodity • Steam inerting under certain conditions • Dilution of water miscible (soluble) flammable and combustible liquids under certain conditions	• Aerosol products • Ammonium nitrate • Compressed and liquified gases in portable cylinders • Flammable and combustible liquids under certain conditions • Liquid and solid oxidizers • Ordinary commodities, plastics, elastomers and rubber • Organic peroxide formulations	• Materials that react explosively with water • Materials that spontaneously ignite or release toxic or flammable gases when in contact with water. • Metals	• Freezes below 32°F (0°C) • Electrically conductive • Water damage potential to commodity • May transport water soluble chemicals • Reactivity with some chemicals	• Low cost • Availability • Environmentally benign • Non-toxic • Superior thermodynamic properties

[1] In the form of droplets or spray

out a containment or drainage system, liquid run-off could be an unwanted by-product of a fire. The liquid run-off could include sprinkler water discharge and released material. Furthermore, even in warehouses not provided with automatic sprinkler systems, the potential for contaminated run-off from fire hose streams applied by fire fighters must be considered. In fact, fire control by manual means will usually require far higher water application rates, and thus more runoff, than with automatic systems. Chapter 6 contains more detail on containment and drainage systems.

7.3. Fire Control, Suppression, and Extinguishing Systems

TABLE 7-3
Sprinkler System Design Criteria References

Commodity Type	NFPA Code or Standard
Aerosol products	30B
Ammonium nitrate	490
Compressed and liquefied gases in portable cylinders	55
Ethylene oxide	560
Flammable and combustible liquids	30
Liquid and solid oxidizers	430
Ordinary commodities, plastics, elastomers and rubber	13, 231 and 231C
Organic peroxide formulations	43B

7.3.3. Fire Extinguishment

Fire extinguishment can be attained with certain commodities by using a sprinkler system that discharges a low expansion foam agent. Low expansion foam is a mixture of foam agents typically at concentrations of 1, 3, or 6 percent with water. When agitated with air it forms an aggregation of small bubbles. The volumetric expansion ratio for low expansion foam is up to 20:1. Foam–water sprinkler systems have been successfully tested with ordinary commodities (see Glossary) and containerized storage of flammable and combustible liquids. The cooling capabilities of foam-water is the same as ordinary water. Additionally, it also acts as a wetting agent with ordinary commodities. Furthermore, foam-water can extinguish burning flammable and combustible liquid pool fires while cooling exposed containers.

Low expansion foam systems generally require higher maintenance and thus offer less reliability than water-only systems. Such systems do offer the advantage (over medium and high expansion foam) of a "water only" discharge after the foam agent has been expended. The sprinkler density also can be designed for a higher than the minimum recommended discharge for a closed-head foam water system, which may improve the probability of fire control.

Medium and high expansion foam agents can be effective in achieving fire extinguishment with certain commodities. Medium and high expansion foam are mixtures of foam agent typically at concentrations of

1½ to 2 percent with water. When agitated with air it forms an aggregation of small bubbles. The volumetric expansion ratio for medium expansion foam is between 20:1 and 200:1. The volumetric expansion for high expansion foam is between 200:1 and 1000:1. When used in conjunction with sprinklers discharging water, high expansion foam can have a synergistic effect with ordinary commodities. Table 7-4 offers some selection parameters associated with low, medium and high expansion foam agents.

TABLE 7-4
Foam Agents

Protection Mechanism	Recommended for Fires Involving	Not Recommended for Fires Involving	Limitations	Advantages
Low Expansion				
• Reduces vapor production with liquid fuels • Separates fuel from air • Cooling • Prevents radiative feedback	• Flammable and combustible liquid spills • Ordinary commodities	• Materials that react with water or foam agent • Metals	• Less effective with pressurized or spill fires • Electrically conductive • Requires clean up after discharge	• Secures fuel from reignition • Acts as wetting agent with ordinary commodities • Non-toxic
Medium and High Expansion				
• Reduces vapor production with liquid fuels • Separates fuel from air • Cools fuel • Prevents radiative feedback • Provides an inert atmosphere of steam	• Flammable and combustible liquids • Ordinary commodities	• Materials that react with water or foam agent • Metals	• Less effective with pressurized or spill fires • Electrically conductive • Requires clean up after discharge • Disorientation and breathing difficulty may result	• Secures fuel from reignition • Acts as wetting agent with ordinary commodities • Synergistic effect when used in conjunction with water-based automatic sprinkler systems • Non-toxic

7.4. Fire Detection and Alarm Systems

Total flooding gaseous agents can also be effective in achieving extinguishment with certain commodities. For these agents to successfully extinguish a fire, an essentially gas-tight enclosure must be provided to maintain the gas concentrations at or above the recommended lower limit. Table 7-5 offers some selection parameters associated with gaseous agents.

Dry chemicals are another class of agents that can be very effective at fire extinguishment with certain commodities. Dry chemical systems are usually used as "local application" systems. However, they can also be used as "total flooding" systems. A total flooding system would typically be used in a small cut-off room that is completely enclosed under certain circumstances. The gaseous and dry chemical agent systems can be used to provide a first line of defense and be used to supplement a water-based sprinkler system. Table 7-6 lists some selection parameters associated with dry chemical agents and identifies these agents by their base component.

Combustible metals require the use of special dry compound agents for extinguishment and are applied manually with either hand scoops or handheld fire extinguishers. These agents are also sold under various proprietary names. Therefore, Table 7-7, which lists some selection parameters associated with these agents, also identifies these agents by their base component.

7.3.4. Fire Extinguishers

Portable fire extinguishers are intended for use against small fires and are available as both handheld and wheeled units. Personnel trained in their proper use must respond to the scene with the appropriate type of fire extinguisher. All of the agents listed in Table 7-1 are available in fire extinguishers. Guidance on their use can be found in NFPA 10, "Standard for Portable Fire Extinguishers."

A list of references for NFPA standards covering specific fire protection equipment systems can be found in Table 7-8.

7.4. Fire Detection and Alarm Systems

In the event of a fire, fire detection and alarm systems allow for fire detection, early warning and evacuation of warehouse occupants and for alerting emergency responders and warehouse management. A fire detection and alarm system may also be utilized to automatically activate devices

TABLE 7-5 Gaseous Agents

Protection Mechanism	Recommended for Fires Involving	Not Recommended for Fires Involving	Limitations	Advantages
Carbon dioxide				
Inerting Cooling (direct application of agent)	• Ordinary commodities • Flammable and combustible liquids • Flammable gases • Materials that do not provide their own oxygen supply • Commodities where liquid runoff due to sprinkler discharge is not desirable • Electrical fires	• Certain metals • Metal hydrides • Oxygen producing material such as cellulose nitrate and oxidizers • Pesticides impregnated on ammonium nitrate fertilizer (oxidizer)	• Concentrations required for extinguishment are lethal • Sealed compartments required for total flooding systems • Agent discharge may adversely affect occupants of compartments due to noise, low visibility and physiological effects	• Readily dispersed • Electrically non-conductive • Self-pressurizing agent
IG-541 [A mixture of nitrogen (52%), argon (40%) and carbon dioxide (8%)]				
Inerting	• Ordinary commodities • Flammable and combustible liquids • Flammable gases • Materials that do not provide their own oxygen supply • Commodities where liquid runoff due to sprinkler discharge is not desirable • Electrical fires	• Certain metals • Metal hydrides • Oxygen producing material such as cellulose nitrate and oxidizers • Pesticides impregnated on ammonium nitrate fertilizer (oxidizer)	• Sealed compartments required for total flooding systems	• Concentrations required for extinguishment are not immediately lethal • Readily dispersed • Electrically non-conductive • Self-pressurizing agent • Environmentally benign
HFC-227ea [Heptafluoropropane]				
Chemically inhibits combustion chain reaction	• Ordinary commodities • Flammable and combustible liquids • Electrical fires	• Metals • Materials that react with heptafluoropropane	• Sealed compartments required for total flooding systems • May generate corrosive by-products when exposed to fire	• Concentrations required for extinguishment are not immediately lethal • Readily dispersed • Electrically non-conductive • Self-pressurizing agent

TABLE 7-6. — Dry Chemical Agents

Type	Protection Mechanism	Recommended for Fires Involving	Not Recommended for Fires Involving	Limitations	Advantages
Urea and potassium bicarbonate / Potassium chloride / Potassium bicarbonate / Sodium bicarbonate	• Chemically inhibits combustion chain reaction • Blocks radiation • Cooling	• Flammable and combustible liquids • Electrical fires	• Ordinary commodities • Metals	• Not effective against fire with reignition potential after discharge completed • Loss of visibility and breathing difficulty during discharge • A 120°F (49°C) temperature limit, otherwise agent can fuse and not flow • Fixed systems not practical for large open warehouses • Clean up of powder after discharge	• Most effective agent for fire knock-down involving flammable and combustible liquids • Non-toxic • Electrically non-conductive
Multipurpose or A B C (monoammonium phosphate)	• Chemically inhibits combustion chain reaction • Blocks radiation • Cooling • Covers ordinary commodities with a metaphosphoric acid coating	• Ordinary commodities • Commodities where liquid runoff due to sprinkler discharge is not desirable • Flammable and combustible liquids • Electrical fires	• Metals • Chlorine producing oxidizers such as calcium hypochlorite	• Not effective against fire with reignition potential after discharge completed • Loss of visibility and breathing difficulty during discharge • 120°F (49°C) temperature limit, otherwise agent can fuse and not flow • Fixed systems not practical for large open warehouses • Clean up of powder after discharge	• Most effective agent for fire knock-down involving flammable and combustible liquids • Non-toxic • Electrically non-conductive

TABLE 7-7
Combustible Metal Agents

Protection Mechanism	Recommended for Fires Involving	Not Recommended for Fires Involving	Limitations	Advantages
Sodium Chloride Base Plus Additive				
• Cooling • Separates fuel from air	• Aluminum • Magnesium • Potassium • Sodium • Sodium-potassium alloy • Titanium • Uranium • Zirconium	• Other metals and commodities not specified.	• Typically not used for automatic systems but applied manually or from a fire extinguisher.	Non-toxic
Sodium Carbonate Base Plus Additives				
• Cooling • Separates fuel from air	• Nickel (Raney) • Sodium	• Other metals and commodities not specified.	• Typically not used for automatic systems but applied manually or from a fire extinguisher.	Non-toxic
Graphite Base Plus Additives				
• Cooling • Separates fuel from air	• Beryllium • Calcium • Lithium • Magnesium • Potassium • Sodium • Sodium-potassium alloy • Titanium • Uranium • Zirconium	• Other metals and commodities not specified.	• Typically not used for automatic systems but applied manually or from a fire extinguisher.	Non-toxic
Copper Powder				
• Cooling • Separates fuel from air	• Aluminum • Lithium • Magnesium	• Other metals and commodities not specified.	• Typically not used for automatic systems but applied manually or from a fire extinguisher.	Non-toxic.

7.4. Fire Detection and Alarm Systems

TABLE 7-8
Fire Protection Equipment Systems

Fire Protection Systems and Related Equipment	NFPA Standard
Carbon Dioxide Extinguishing Systems	12
Centrifugal Fire Pumps	20
Clean Agent Fire Extinguishing Systems	2001
Closed-Head Foam-Water Sprinkler Systems	16A
Deluge Foam-Water Sprinkler and Foam-Water Spray Systems	16
Dry Chemical Extinguishing Systems	17
Halon 1301 Fire Extinguishing Systems	12A *
Installation of Sprinkler Systems	13
Medium- and High-Expansion Foam Systems	11A
Portable Fire Extinguishers	10
Private Fire Service Mains and Their Appurtenances	24
Standpipe and Hose Systems	14
Water Tanks for Private Fire Protection	22

* Typically no longer installed due to Montreal Protocol. However, systems are still in service

upon "alarm" conditions, such as automatic closing fire doors and storm water drain valves. Alarm systems can also provide supervisory status of fire protection equipment and watchman service. Some fire alarm control panels can also function as a automatic release system to activate various extinguishing systems including deluge, preaction, and total flooding systems.

The various types of fire alarm systems can include those that are "local," "proprietary," or those connected to a remote or central station.

A local fire alarm system is installed within the building it protects. The alarm signals are not transmitted to a remote location. These systems usually have bells, horns and strobe lights to notify the occupants. The fire department is dependent upon the occupants for notification of a fire.

A proprietary fire alarm system can protect one or more buildings under one ownership. The fire alarm systems are monitored 24 hours a day at a central monitoring location which comes under the same ownership as the building(s) being protected. The central monitoring station may be located locally or remote from the building(s) being protected. Fire department notification is provided by the central monitoring station.

Remote and central station fire alarm systems are those monitored by a local government agency or private alarm company respectively.

A wide variety of sensing and initiating devices are available that can detect heat, smoke, radiant energy, and fire gas products. These devices should be selected based upon the type of fire anticipated. Additionally, sprinkler system water flow, activation of other extinguishing systems, and manual fire alarm "pull stations" can also be used to initiate an alarm signal.

Annunciators and display panels should be selected after considering the warehouse environment. For example, the ambient noise levels may effect the type of audible annunciators that are selected.

Central station services should be listed or approved by a nationally recognized laboratory.

For guidance on fire detection and alarm system design, installation, testing and maintenance, see NFPA 72. For additional discussion on inspection, testing, and maintenance programs see Chapter 8.

References

National Fire Protection Association, "Standard for Portable Fire Extinguishers." NFPA 10, Quincy, MA, 1994
National Fire Protection Association, "Standard for Medium- and High-Expansion Foam Systems." NFPA 11A, Quincy, MA, 1994
National Fire Protection Association, "Standard on Carbon Dioxide Extinguishing Systems." NFPA 12, Quincy, MA, 1993
National Fire Protection Association, "Standard for the Installation of Sprinkler Systems." NFPA 13." Quincy, MA, 1996
National Fire Protection Association, "Standard for the Installation of Standpipe and Hose Systems." NFPA 14, Quincy, MA, 1996
National Fire Protection Association, "Standard for the Installation of Deluge Foam-Water Sprinkler and Foam-Water Spray Systems." NFPA 16, Quincy, MA, 1995
National Fire Protection Association, "Standard for the Installation of Closed-Head Foam-Water Sprinkler Systems." NFPA 16A, Quincy, MA, 1994
National Fire Protection Association, "Standard for Dry Chemical Extinguishing Systems." NFPA 17, Quincy, MA, 1994
National Fire Protection Association, "Standard for the Installation of Centrifugal Fire Pumps." NFPA 20, Quincy, MA, 1996
National Fire Protection Association, "Standard for Water Tanks for Private Fire Protection." NFPA 22, Quincy, MA, 1996
National Fire Protection Association, "Standard for the Installation of Private Fire Service Mains and Their Appurtenances." NFPA 24, Quincy, MA, 1995

National Fire Protection Association, "Flammable and Combustible Liquids Code." NFPA 30, Quincy, MA, 1996
National Fire Protection Association, "Code for the Manufacture and Storage of Aerosol Products." NFPA 30B, Quincy, MA, 1994
National Fire Protection Association, "Code for the Storage of Organic Peroxide Formulations." NFPA 43B, Quincy, MA, 1993
National Fire Protection Association, "Code for the Storage of Pesticides." NFPA 43D, Quincy, MA, 1994
National Fire Protection Association, "Standard for the Storage, Use, and Handling of Compressed and Liquefied Gases in Portable Cylinders." NFPA 55, Quincy, MA, 1993.
National Fire Protection Association, "National Fire Alarm Code." NFPA 72, Quincy, MA, 1996
National Fire Protection Association, "Standard for General Storage." NFPA 231, Quincy, MA, 1995
National Fire Protection Association, "Standard for Rack Storage of Materials." NFPA 231C, Quincy, MA, 1995
National Fire Protection Association, "Code for the Storage of Liquid and Solid Oxidizers." NFPA 430, Quincy, MA, 1995
National Fire Protection Association, "Code for the Storage of Ammonium Nitrate." NFPA 490, Quincy, MA, 1993
National Fire Protection Association, "Standard for the Storage, Handling, and Use of Ethylene Oxide for Sterilization and Fumigation." NFPA 560, Quincy, MA, 1995
National Fire Protection Association, "Standard on Clean Agent Fire Extinguishing Systems." NFPA 2001, Quincy, MA, 1996
National Fire Protection Association, "Fire Protection Handbook." Eighteenth Edition, Quincy, MA, 1997

Additional Reading

Factory Mutual System, "Approval Guide 1995." Norwood, MA, 1996
National Fire Protection Association, "Fire Prevention Code." NFPA 1, Quincy, MA, 1997
National Fire Protection Association, "Guide to the Fire Safety Concepts Tree." NFPA 550, Quincy, MA, 1995
Underwriters Laboratories, "Fire Protection Equipment 1996." Northbrook, IL, 1996

8 | Inspection, Testing, and Maintenance Programs

8.1. Synopsis

Reliability of those systems and building components that contribute to safety or environmental and property protection can be maximized through comprehensive inspection and test and maintenance programs. Because many of these systems and components are not exercised except in emergencies, breakdowns or failures may not be readily detectable other than through inspections and tests.

8.2. Inspection and Test Programs

8.2.1. Program Objectives

The objective of an inspection and test program is to assure the reliable performance of critical equipment and construction features. Since most safety related equipment and construction features remain idle until needed under emergency conditions, their condition and readiness often cannot be verified through normal use. Inspections and tests provide essential verification that they will perform effectively when needed.

8.2.2. Critical Equipment and Construction Features

Equipment and building components should be included in an inspection and test program if their failure to perform is not readily apparent during the course of day-to-day operations, and if their failure could contribute to injury, environmental damage or significant property loss. Some examples are:

- Fire detection, alarm, control and suppression systems.
- Personal protection equipment (emergency respirators, decontamination suits, etc.).
- Emergency decontamination and clean up kits.
- Refrigeration system failure, high temperature alarms.
- Fire walls and fire doors.
- Storage racks, in which structural damage may be difficult to detect.
- Industrial truck safety devices.
- Roof integrity.
- Floor condition, where impermeability is required.

8.2.3. Inspection and Test Program Elements

Inspection and test programs should include:

- policy, purpose and scope statements covering the overall program.
- program responsibilities by name and title.
- lists identifying all equipment and components subject to the program, and those that are critical to safety.
- inspection and test procedures, or references to explicit procedures.
- detailed inspection and test records.
- a system for periodic auditing and follow-up.

The Policy Statement should reflect the commitment of management to a high standard of readiness of critical equipment and should give high priority to the inspection and test program. The Purpose Statement should explain the unique importance of the program to safety, the environment and property protection. The Scope Statement should describe the classes of equipment and components included in the program.

Program responsibilities should be assigned both to the member of management in overall charge, and to the group or individuals who will perform the inspections and tests. Management responsibilities include assuring that the program is carried out as designed, that persons performing the inspections and tests are qualified by training and experience, and that sufficient resources are allocated for the program.

Equipment lists must be tailored to the site. The broad classes of equipment and components listed above should be expanded to include each system and device, down to the individual fire extinguisher, respira-

8.2. Inspection and Test Programs

tor and safety shower. Only through this degree of detail can the possibility of overlooking a key inspection be minimized.

Inspection procedures should be developed based on regulatory requirements, consensus standards (for example, National Fire Protection Association codes and standards. See Table 8.1), manufacturers' recommendations, and experience. Procedures should include what is to be done, its purpose, what is an acceptable result, and the frequency at which it is to be performed. A log or check list can be utilized to assure accuracy during maintenance.

Records should include, what was done, when, by whom and the results of the inspection or test.

The follow up system should include a "tickler" file which is kept open until any identified deficiencies are corrected and the equipment's readiness is verified. Deficiencies in critical equipment may warrant shutting the equipment down until repairs are made.

Some records must be retained for specified periods by law. In any case, records should be reviewed periodically to identify repetitive failures which might suggest the need for revised maintenance or design, or more frequent inspections and tests. Conversely, continuously satisfactory results might justify increasing the interval between inspections and tests, subject to statutory requirements. Reliability-based maintenance techniques are valuable tools in this area.

8.2.4. Maintenance

Maintenance programs both prevent failures and prolong the useful life of equipment. In the case of safety—critical equipment and components, failure prevention is of primary concern, however prolonging useful life will also reduce failure frequency and thus contribute to overall reliability.

Preventive maintenance programs should include policy, purpose and scope statements and other elements similar to those described earlier for inspection and test programs.

Where maintenance or repair of certain safety - critical equipment may require that it be taken out of service, extra precautions may be needed, for example:

- Fire watches or extra patrols may be needed.
- Welding and other spark or flame producing activities, such as industrial truck traffic, should be suspended.
- Fire department, central station office and insurance company should be alerted to the impairment.

- Temporary water supply connections may be made to bridge impaired components.
- Extra hazardous materials storage could be eliminated or reduced.
- Replacement units should be provided when fire extinguishers are removed for testing or service.

8.2.5 Maintenance Procedures

Table 8.1 lists sources of information on maintenance of many types of safety and emergency equipment and systems.

References

American National Standards Institute "Emergency Eyewash and Shower Equipment." ANSI Z 358.1 - 1990, New York, NY, 1990.
National Fire Protection Association, "Standard for Portable Fire Extinguishers." NFPA 10, Quincy, MA, 1994.
National Fire Protection Association, "Standard for Medium- and High-Expansion Foam Systems." NFPA 11A, Quincy, MA, 1994.
National Fire Protection Association, "Standard on Carbon Dioxide Extinguishing Systems." NFPA 12, Quincy, MA, 1993.
National Fire Protection Association, "Standard for Dry Chemical Extinguishing Systems." NFPA 17, Quincy, MA, 1994.
National Fire Protection Association, "Standard for the Inspection, Testing, and Maintenance of Water-Based Fire Protection Systems." NFPA 25, Quincy, MA, 1995.
National Fire Protection Association, "Guide for Venting of Deflagrations." NFPA 68, Quincy, MA, 1994.
National Fire Protection Association, "National Electric Code." NFPA 70, Quincy, MA, 1996.
National Fire Protection Association, "National Fire Alarm Code." NFPA 72, Quincy, MA, 1996.
National Fire Protection Association, "Standard for Fire Doors and Fire Windows." NFPA 80, Quincy, MA, 1995.
National Fire Protection Association, "Life Safety Code." NFPA 101, Quincy, MA, 1997.
National Fire Protection Association, "Standard for Emergency and Power Supply Systems" NFPA 110, Quincy, MA, 1996.
National Fire Protection Association, "Standard on Stored Electrical Energy Emergency and Standby Power Systems." NFPA 111, Quincy, MA, 1996.
National Fire Protection Association, "Guide for Smoke and Heat Venting." NFPA 204M, Quincy, MA, 1991.
National Fire Protection Association, "Fire Safety Standard for Powered Industrial Trucks Including Type Designations, Areas of Use, Conversion, Maintenance, and Operation." NFPA 505, Quincy, MA, 1996.
National Fire Protection Association, "Standard for the Installation of Lighting Protection Systems." NFPA 780, Quincy, MA, 1995.

8.2. Inspection and Test Programs

TABLE 8.1
Inspection, Testing, and Maintenance Guidelines

Equipment or System	Code, Standard or Source
Carbon Dioxide Extinguishing Systems	NFPA 12 *
Centrifugal Fire Pumps	NFPA 25
Clean Agent Fire Extinguishing Systems	NFPA 2001 *
Closed-Head Foam-Water Sprinkler Systems	NFPA 25
Communication Systems	Manufacturer's Recommendations
Deflagration Vents	NFPA 68
Deluge Foam-Water Sprinkler Systems	NFPA 25
Dry Chemical Extinguishing Systems	NFPA 17 *
Electrical Equipment	NFPA 70
Emergency and Standby Power Systems	NFPA 110 and 111
Emergency Lighting	NFPA 101
Emergency Spill Equipment	Manufacturer's Recommendations
Eye Wash Stations	ANSI Z 358.1
Fire Alarm Systems	NFPA 72
Fire Doors and Windows	NFPA 80
Floors	Manufacturer's Recommendations
Heat and Smoke Vents	NFPA 204M
Lightning Protection	NFPA 780
Medium- and High-Expansion Foam Systems	NFPA 11A *
Personal Protective Equipment	Manufacturer's Recommendations
Portable Fire Extinguishers	NFPA 10 *
Powered Industrial Trucks	NFPA 505
Powered Ventilation systems	Manufacturer's Recommendations
Private Fire Service Mains	NFPA 25
Refrigeration Systems	Manufacturer's Recommendations
Roofs & Roof Drains	Manufacturer's Recommendations
Security Fixtures	Manufacturer's Recommendations
Sprinkler Systems	NFPA 25
Standpipe and Hose Systems	NFPA 25
Water Tanks for Private Fire Protection	NFPA 25

* Contains recharging procedures.

National Fire Protection Association, "Standard on Clean Agent Fire Extinguishing Systems." NFPA 2001, Quincy, MA, 1996.

National Safety Council, *Accident Prevention Manual for Industrial Operations Administration and Programs*, 10th ed. Itasca, IL, 1992.

Additional Reading

National Fire Protection Association, "Standard for the Installation of Sprinkler Systems." NFPA 13, Quincy, MA, 1996.

National Fire Protection Association, "Standard for the Installation of Standpipe and Hose Systems." NFPA 14, Quincy, MA, 1996.

National Fire Protection Association, "Standard for the Installation of Deluge Foam-Water Sprinkler and Foam-Water Spray Systems." NFPA 16, Quincy, MA, 1995.

National Fire Protection Association, "Standard for the Installation of Closed-Head Foam-Water Sprinkler Systems." NFPA 16A, Quincy, MA, 1994.

National Fire Protection Association, "Standard for the Installation of Centrifugal Fire Pumps." NFPA 20, Quincy, MA, 1996.

National Fire Protection Association, "Standard for Water Tanks for Private Fire Protection." NFPA 22, Quincy, MA, 1996.

National Fire Protection Association, "Standard for the Installation of Private Fire Service Mains and Their Appurtenances." NFPA 24, Quincy, MA, 1995.

9. Emergency Planning

9.1. Synopsis

A comprehensive warehouse risk management strategy must include an emergency plan that addresses emergency and potential loss scenarios. This plan should provide a means to evaluate and minimize exposure to employees, surrounding population, surrounding environment, the warehouse, and business continuity. Properly implemented, the plan could make the difference between a minor incident and a catastrophe. The plan should be reevaluated on a regular basis to account for changes that may occur after the plan's initial development.

9.2. Loss Scenarios

Potential loss scenarios include fire, explosion, chemical release, natural peril occurrences, riot and civil commotion, arson, loss of utilities, or a fire, explosion, or a chemical release at a nearby facility. Chemical releases might result from a spill or container failure, or as a consequence of fire, natural peril, or other event, and might involve the chemicals themselves, products of chemical reactions or decomposition, or in the case of fire, products of combustion.

9.3. Plan Objectives

The objective of the plan is to protect employees, the surrounding population and the environment from the consequences of accidents, and to minimize property damage and business interruption. The process of developing an emergency plan may also reveal opportunities to further reduce risk through changes in storage arrangement, protection, design or procedures.

9.3.1. Employees

The emergency plan should be designed to prevent injury to employees and other building occupants while minimizing exposures to the community. The plan should include a strategy to evacuate all warehouse occupants in the event of an incident that could adversely affect their health or safety. Emergency medical response and care should also be considered.

9.3.2. Surrounding Population

Minimizing the risk to the surrounding population may require provisions for evacuation. Community evacuation planning is primarily the responsibility of local authorities and emergency responders in cooperation with warehouse management.

9.3.3. Environment

Particular attention should be given where facilities are located in highly sensitive environmental areas, adjacent to bodies of water, aquifers, or on porous soil. Water and porous soils allow for greater transportability of released chemicals. Chapter 5 contains additional information on environmental considerations.

9.3.4. Property Protection and Business Interruption

Emergency planning should consider the possibility of property damage and business interruption resulting from contamination of building or contents by released chemicals as well as by fire, windstorm, flood or other events.

9.4. Plan Development

Emergency plan development and reevaluation is a cooperative effort involving management, operations, distribution, maintenance, ancillary services, emergency response and governmental agencies. The planning

9.5. Plan Elements

team should review other pertinent company and community plans to assure consistency.

Emergency planning for new facilities should begin prior to construction and occupancy when identified design problems may be more easily corrected. Existing facilities should reevaluate their plans periodically, since changes in types of materials stored, storage arrangements and the surrounding community can have a significant effect on potential loss scenarios.

The Chemical Manufacturers Association publication, *Site Emergency Response Planning Guidebook*, provides additional guidance on the emergency planning process.

9.5. Plan Elements

9.5.1. Policy Statement

A policy statement should outline the risk management philosophy, commitment, and overall responsibilities and objectives associated with the emergency plan.

9.5.2. Scope and Objectives

This plan's scope and objectives provides:

- Plan objectives by priority.
- Loss scenarios anticipated.
- Individuals responsible for plan implementation.
- Areas of responsibility of plan participants.
- Intended audience for the plan.
- Individuals responsible for plan development.
- Effective dates.

9.5.3. Pre-Incident Planning

9.5.3.1. Risk Assessment

The initial step in pre-incident planning is a risk assessment, based on an inventory of all material stored at the warehouse. This information should be readily available as part of the Hazard Communication program outlined in Chapter 4. The properties and hazards of stored materials should be determined using the material safety data sheets and commodity hazard information as discussed in Chapter 2. Special hazards should be identified for use in developing emergency response plans.

The possible loss scenarios, including events originating both internally and externally, should be explored. An assessment should then be made of the controls and protection features provided and their mitigative effect on each loss scenario. Failures of controls and protection features should then be assumed to determine the "worst-case" and other, more credible, loss scenarios. Finally, the likelihood and degree of risk associated with each loss scenario should be developed.

The risk assessment should be reviewed and reassessed periodically and when major changes take place in the warehouse.

Risk potential from outside of the warehouse, such as exposure to nearby high hazard operations should also be evaluated. Mitigating features, such as separation distance, fire walls, emergency response plans, and mutual aid programs should be considered in this evaluation.

9.5.3.2. Emergency Responders

A timely and effective response will be necessary to mitigate the effects of "worst-case" and other, more credible, loss scenarios.

A designated emergency response coordinator in overall charge of the situation should be specified. Responsibility may at some point during an emergency be transferred to another individual, for example, from warehouse management to the fire chief of the local fire department upon their arrival on the scene.

An emergency response team, under the control of the emergency response coordinator, should be assembled with a clearly defined chain of command. The members of this team should be selected and trained for specific mitigative and communication tasks. Specific duties might include spill cleanup, first aid, employee evacuation and headcount, incipient fire response, fire pump/sprinkler control valve/fire door monitoring, firefighting, fire/emergency responder communications, and rescue.

Outside emergency responders should visit the facility as part of their pre-incident training. This training should include a knowledge of warehouse layout, storage arrangement and associated commodity hazards, protection features, and the emergency plan. Pre-incident planning should include advising local hospitals on the hazards associated with stored chemicals.

9.5.3.3. Resource Determination

Based upon the loss scenarios defined, the plan should prescribe the manpower and equipment resources needed as outlined in Section 9.6.

9.5. Plan Elements

This section suggests some equipment that should be maintained at the warehouse for relatively small chemical spills, including personal protective equipment. Additionally, Chapter 7 provides information on the types of fire extinguishers that would be useful against incipient fires.

For incidents beyond the capabilities of the resources at the warehouse, appropriate outside agencies such as public emergency responders, environmental and emergency management agencies, remediation firms and mutual aid groups should be included in the plan. The need for an off-site emergency coordination and communication control center should also be considered.

When the incident might affect the surrounding community, the appropriate authorities should be notified immediately. This is particularly relevant if the population proximity, density, or sensitivity presents a high risk. See also Chapter 5.

9.5.3.4. Evacuation

Employees should be instructed and drilled on when and how to evacuate and where to assemble after leaving the warehouse. An evacuation plan should include a person or persons assigned to ensure that all employees have vacated the warehouse and have been accounted for. Employee notification that an emergency condition exists could be done through the use of a public address system or the warehouse fire alarm system.

9.5.3.5. Emergency Notification Procedures

A system of emergency notification should be developed that is both internal and external to the warehouse operation. The system should include a list of contact names and telephone numbers including the emergency response coordinator, members of the emergency response team, and company officials. This list should provide for 24 hour contact with alternate names for each position. It should also state under what circumstances these individuals should be contacted.

A similar list of contacts should also be prepared for outside agencies. The circumstances under which these agencies should be contacted should be specified, along with their role and responsibility.

9.5.3.6. Site Security

The plan should include measures to secure the site and prevent unauthorized access following an incident. This is necessary not only for safeguarding employees and contractors, but also to prevent tampering with the area before a loss investigation can be conducted.

9.5.3.7. Training and Evaluation

Training should occur during the introduction of the emergency plan and when significant changes are made. Testing and evaluation of the emergency plan should be conducted by performing periodic exercises.

9.5.3.8. Remediation

The primary remediation activity should be control of atmospheric releases or spills or run-off within the defined containment area, and to make certain that hazardous materials do not enter the sewer system, porous soil, or surface or ground water. This section of the plan should address not only the initial containment activity, but also post-incident clean-up, waste disposal, and environmental restoration. Safe clean-up methods should be outlined in the plan.

9.5.3.9. Public Relations

A public relations program should be an integral part of a comprehensive emergency plan. This program allows a designated spokesperson to keep the media, public, and interested government agencies informed during and following an incident. The emergency plan should include a list of media contacts and designate a media staging area before an incident. The plan should also address the expectation that a large number of calls may come into the facility or off-site control center following a major incident.

9.5.4. Incident Response

9.5.4.1. Situation Assessment

If a situation arises that has or could evolve into a loss, an assessment must be conducted to determine the level of risk. Of concern in this assessment would be the exposure to employees, nearby communities, and the environment. This assessment should be an ongoing process requiring constant monitoring. To assure appropriate responses to each situation by the on-site coordinator, use of a checklist or decision tree is suggested.

9.5.4.2. Emergency Plan Activation

Once a situation assessment has been made and the appropriate response has been determined, the emergency response coordinator should assume responsibility. At this point, sub-elements of the emergency plan should be activated on an "as-needed" basis, for example, emergency notification, resource coordination, evacuation procedures, site security, remediation, and public relations. The primary concerns during this period should be the safeguarding of lives, emergency medical treatment, and the containment of any chemical releases.

9.5.5. Post Incident Activity

After the incident has been declared under control and acute emergency conditions no longer exist, post-incident activity and stabilization should commence. Activity during this period may include cleanup and disposal, accident investigation/incident reconstruction, site restoration, and emergency plan evaluation.

9.6. Emergency Spill Response

While safe warehouse operations are the first defense against chemical releases, the emergency plan should include provisions for handling spills.

9.6.1. Planning

The spill response portion of the plan should include information on:

- identification of appropriate action to be taken for potential releases
- selection and training of a lead employee and spill response team for each shift
- purchase and assembly of personal protective equipment and spill control supplies
- awareness training of all warehouse personnel, and
- plan drills.

An emergency spill response plan should reflect the operations, materials and quantities which are specific to that facility. Inventory should be maintained of all chemicals, and the quantity on site of each. Information on the hazardous characteristics of each chemical may be found on Material Safety Data Sheets (MSDSs), which are required by law to accompany the first chemical shipment. Appropriate responses will depend upon an evaluation of the hazards posed from release of the chemicals specific to each facility.

Planners should identify possibilities for emergencies involving the chemicals in that facility. Each such possibility should be evaluated and prioritized. The plan should evolve out of this process.

Once the hazardous nature of the facility's inventory have been assessed, management should evaluate its capacity to respond to each potential type of incident. Deficiencies identified in the response plan should be addressed as the plan is developed.

Chemical warehouse operators should identify a spill response team who take immediate action in the event of a spill. The team should be headed by a lead employee, usually a foreman or plant manager, who will supervise the response.

Appropriate spill control supplies should be assembled on a cart or pallet which can be quickly delivered to the site of a spill. Typical spill clean up supplies are listed in Table 9-1.

TABLE 9-1
Typical Spill Cart Supplies

Absorbents: • Pillows Particles • Sheets Booms • Rolls Pads		Chemical suits; flash fire protective clothing for large flammable liquid spills
Solidifiers and Neutralizers		Goggles
Portable Fire Extinguishers • AFFF (foam) • Purple K • ABC Dry Chemical		Respirators [3] • Organic Vapor • Acid • Chlorine
Scoop		Self-Contained Breathing Apparatus[4]
Shovels		Grounding cable with clamps
Air-powered vacuum [1]		Gloves
Containers, oversized drums, pails, etc.		Drum plugs
Barricade tape/flagging		Epoxy putty
Plastic dip strip & color change indicators		Duct tape
Detection instruments to determine extent of spill hazard [2]		Magnetic drain guards

[1] Air powered vacuums must be approved for use with hazardous chemicals. [2] Air sampling to determine concentrations of hazardous chemicals requires specialized training.

[3] Respirator usage is covered in Section 4.5. Respirators are not required for all spills.

[4] SCBA requires specialized training.

9.6. Emergency Spill Response

9.6.2 Responding to a Hazardous Material Spill

Responsibilities of the lead employee and spill response team include:

- Evacuating the area. Only the response team should remain in the area and only if appropriate safety measures are taken. Verbal evacuation instructions should be given to employees and eventually the area should be roped off and posted. The area should not be left unguarded.
- Wearing personal protective equipment (PPE).
- Providing first aid to affected individuals. The lead employee must be familiar with the appropriate first aid assistance for each chemical present at the warehouse, and know when to seek emergency medical assistance.
- Shutting down equipment and halting operations within the immediate area of the spill.
- Shutting off all sources of ignition, such as pilot lights and industrial truck operations in potential fire or explosion situations.
- Activating ventilation equipment when vapors, gases and dusts need to be evacuated from the area.
- Mobilizing spill control supplies and containing the spill. Further leakage of the chemical should be contained. Once contained, the spill should be covered. An absorbent material is often used to cover a liquid spill. Plastic tarpaulins are often used if the spill is dry material.
- Notifying owners and/or managers of any accidental spill including: the location of the event, the nature of the occurrence, the extent of the damages (including injuries and exposures to individuals, fire and environmental contamination), and the name and the quantity of the chemical involved.

9.6.3. Cleanup

The lead employee, or other appropriately qualified person, should supervise cleanup of the spill. The persons involved in this cleanup must be thoroughly trained, protected and supervised. Each person present should wear appropriate PPE, including respiratory equipment, if necessary. The spill should be approached from the edges only. At no time should any employee be allowed to work inside a spill area. Spilled material and any absorbents used should be transferred into properly identified containers using scoops and shovels. If spills create a fire or explosion hazard, only approved vacuums and other powered equipment should be used. Disposal should be handled as a hazardous waste, where appropriate.

Under some circumstances, it may be appropriate to contract with a third-party firm which handles emergency response. The plan should determine which, if any, third party firms are to be used, what qualifications are required, under what circumstances they are called, and who should call.

9.6.4. Reporting

Immediate notification of a spill should be made to owners, managers and appropriate emergency response agencies.

Table 9-2 lists the U.S. Federal laws which address the notification of regulatory agencies in the event of a chemical release.

The "Reportable Quantity" of a chemical release is established by Federal law as the threshold volume of each specific chemical, which, if exceeded must be reported to local, state and Federal authorities. Notification need not be made if the spills are confined within the boundary of the facility. Table 9-3 reflects reportable quantities of representative chemicals which have been referenced elsewhere in this document.

- Telephone notification should be made to the community emergency coordinator for the local emergency planning committee for any area likely to be affected by the release. If no local emergency planning committee exists, notification should be made to relevant local emergency response personnel such as the fire department, police department, hazmat team, sanitary district, etc.
- Telephone notification should be made to the state emergency response commission of any state likely to be affected by the release.
- Telephone notification should be made to the National Response Center in Washington, DC at 1-800-424-8802.

TABLE 9-2
Laws Addressing Chemical Releases

CERCLA	40 CFR 302
Clean Water Act	40 CFR 110 & 112
EPCRA (SARA)	40 CFR 355
OSHA-HAZWOPER	29 CFR 1910.120
NCP	40 CFR 300
RCRA	40 CFR 262

9.6. Emergency Spill Response

TABLE 9-3
Reportable Quantities of Releases of Representative Chemicals

Hazardous Substance	Reportable Quantity pounds (kg)
Acetone	5000 (11,000)
Acrylonitrile	100 (220)
Ammonia	100 (220)
Calcium Hypochlorite	10 (22)
Carbaryl	100 (220)
Diazinon	1 (2.2)
Ethyl Ether	100 (220)
Source: 40 CFR 302 CERCLA	

- Information given in the initial telephone notification should be accurate to the extent known at the time of the call and should include:
  - ✓ identification of the chemical involved in the release;
  - ✓ indication of whether the release involved an "extremely hazardous chemical";
  - ✓ an estimate of the volume released into the environment;
  - ✓ the time and duration of the release;
  - ✓ the medium/media into which the release occurred;
  - ✓ any known acute or chronic health risks associated with the emergency and where appropriate, advice regarding needed medical attention;
  - ✓ if appropriate, proper precautions to take (such as evacuation);
  - ✓ name and telephone numbers of contacts.
- Written notification is required for the agencies previously contacted.
- This should include information which updates the telephone notification, including:
  - ✓ actions taken to respond to and contain the release;
  - ✓ any known or anticipated acute or chronic health risks associated with the release, and;
  - ✓ where appropriate, advice regarding medical attention necessary for exposed individuals.

- Local, state and federal agencies may follow-up the initial phone contact with a written request for further information which may not have been available immediately following the incident.
- A written report will not only serve to comply with regulatory requirements, but will assist warehouse management in their efforts to assess the cause of an incident and to prevent future incidents.

All accidents should be investigated to determine the cause and to reduce the possibility that such an incident could be repeated.

Copies of analytical reports, manifests, third party reports, press releases, newspaper clippings, and all other correspondence should be kept on file. These documents may be needed to satisfy regulatory inquiry.

Employees should be cautioned not to give statements or opinions to unknown investigators. Only authorized persons should give statements.

9.6.5. Public Response

While most spill situations do not affect the community, a plan should, nevertheless, provide for this emergency.

The possibility of a chemical release of sufficient magnitude to require evacuation of a community is remote but should be addressed.

For less severe situations, such as where neighbors are affected by odors or observe emergency operations, the plan should address neighborhood notification. Information to the community should be prompt and honest.

9.7. Regulations and Resources

The following regulations and resources may be useful in planning emergency response activities:

9.7.1. U.S. Regulations

9.7.1.1. SARA Title III
Title III of the Superfund Amendment and Reauthorization Act (SARA), also known as The Emergency Planning and Community Right-To-Know Act (EPCRA), 40 CFR 350, 355, 370 and 372, requires each state to appoint a State Emergency Response Commission (SERC). Each state is

9.7. Regulations and Resources

divided into Emergency Planning Districts that are covered by Local Emergency Planning Committees (LEPC). These committees are composed of various interests; such as health officials, emergency responders, community groups, media representatives, and chemical industry representatives. Facilities which store chemicals which are defined in the law as extremely hazardous in quantities in excess of established threshold limits must provide MSDS's, the name of the emergency planning coordinator for the warehouse, and other information as needed to the SERC, LEPC and local fire department. This information will be used in the community planning process.

9.7.1.2. EPA Accident Release Prevention Requirements and Risk Management Programs

Structured as a companion rule to the OSHA Process Safety Management rule, the EPA's Accident Release Prevention Requirements and Risk Management Program, 40 CFR 68—Clean Air Act, seeks to reduce the frequency and severity of chemical accidents in order to protect the surrounding community and the environment. Facilities which manufacture, use, store or otherwise handle regulated substances in excess of specific threshold limits are regulated under this rule. The Risk Management Program requires that these facilities assess and understand the hazards of the warehouse, to take action to prevent chemical releases, to mitigate any accidental releases, and to provide information to the public regarding the nature of the risks of the warehouse.

9.7.1.3. OSHA Process Safety Management

The OSHA Process Safety Management (PSM) Standard, 29 CFR 1910.119, is companion legislation to the EPA Risk Management Program. While the EPA legislation addresses accidents which affect the community and the environment outside the warehouse, PSM addresses harmful chemical releases which could affect workers. Companies which store regulated substances in excess of established threshold limits may be covered, though some exemption is permitted where conditions such as sealed containers at atmospheric pressure exist. Retail facilities are specifically excluded. The rule requires firms having covered processes to develop process safety information, conduct a hazard assessment, develop written processes for operations, maintenance, and management of change, investigate accidents, and train workers on emergency procedures.

9.7.1.4. OSHA Employee Emergency Plans and Fire Prevention Plans and Emergency Procedures

Facilities that handle regulated substances and that are subject to emergency response operations are regulated under 29 CFR 1910.38. This regulation outlines guidelines for planning and conducting emergency operations including pre-emergency planning, personnel roles, notification, evacuation, clean up, emergency medical treatment and first aid, and personal protective equipment.

29 CFR 1910.38 also provides requirements for procedures to follow during fire and other emergencies, including: emergency escape procedures and routes, shut down procedures, procedures for accounting for employees after evacuation, rescue and medical duties, and emergency notification procedures. Such plans are required to include a list of the major fire hazards in the workplace, storage and handling procedures, names and titles of maintenance personnel who handle fire control or extinguishing equipment, and also those responsible for control of fuel source hazards.

9.7.1.5. OSHA Hazard Communication Standard

The Hazard Communication Standard under 29 CFR 1910.1200 requires employees to be knowledgeable in the hazards of the stored materials. (See also Chapter 4.)

9.7.1.6. Resource Conservation and Recovery Act (RCRA)

Facilities which treat, store or dispose of hazardous wastes must, under 40 CFR 264 and 265, develop emergency plans to minimize hazards to human health and the environment in the event of fire, explosion, or chemical release into the environment. A detailed plan of action is required, which considers such events and details arrangements with local police, fire, hospital, emergency response teams. The plan should list the name of the emergency coordinator(s), emergency equipment, and detail the evacuation plan.

9.7.2. CMA Responsible Care Program

The Chemical Manufacturers Association (CMA) Responsible Care Program is designed to provide CMA members with guidance regarding safety, health, and environmental quality. The program includes six codes of practice covering Community Awareness and Emergency Response (CAER), Pollution Prevention, Process Safety, Distribution, Employee Health and Safety, and Product Stewardship.

References

Chemical Manufacturers Association, *Site Emergency Response Planning Guidebook*, January, 1992.
29 Code of Federal Regulations, Part 1910.38, "Employee Emergency Plans and Fire Prevention Plans."
29 Code of Federal Regulations, Part 1910.119, "Process Safety Management of Highly Hazardous Chemicals."
29 Code of Federal Regulations, Part 1910.120, "Hazardous Waste Operations and Emergency Response."
29 Code of Federal Regulations, Part 1910.1200, "Hazard Communication Standard."
40 Code of Federal Regulations, Part 68, "Clean Air Act."
40 Code of Federal Regulations, Parts 110 and 112, "Clean Water Act."
40 Code of Federal Regulations, Parts 262, 264 and 265, "Resource Conservation and Recovery Act."
40 Code of Federal Regulations, Part 300, "National Oil and Hazardous Substances Pollution Contingency Plan."
40 Code of Federal Regulations, Part 302, "Comprehensive Environmental Response, Compensation and Liability Act."
40 Code of Federal Regulations, Parts 350, 355, 370, and 372, "Superfund Amendments and Reauthorization Act (SARA Title III) /Emergency and Community Right to Know Act (EPCRA)."

Additional Reading.

British Columbia Ministry of Environment, Lands and Parks, Environmental Emergency Services Branch, The Environmental Protection Compendium, "Guidelines for Industry Emergency Response Contingency Plans." March, 1992.
Chemical Manufacturers Association, *Safe Warehousing of Chemicals*. Arlington, VA, 1989.
Donovan, "Chemical Companies Formulate Proactive Emergency Plans." *Safety and Health*, April 1992, pp. 34–38.
Federal Emergency Management Agency, "Hazardous Materials Exercise Evaluation Methodology (HM-EEM) Manual." February, 1992.
Indelicato, Gregory J., and Clark, Michael A., "Preparing for Emergency Spill Response." *Pollution Engineering*, 27(1), 56–58, 1995.
LeBar, Gregg, "Planning for the Worst." *Occupational Hazards*, 57(6), 52–55, 1995.
National Fire Protection Association, "Recommended Practice for Pre-Incident Planning for Warehouse Occupancies." NFPA 1420, Quincy, MA, 1993.
Ray, E.R., CSP, WSO-CSM, REP, "Emergency Response Teams—Build One Now!" *Industrial Fire World*, 6(3), 24–27, 1991.
U.S. Environmental Protection Agency, Federal Emergency Management Agency, U.S. Department of Transportation, "Training Guidance for Hazards Analysis: Emergency Planning for Extremely Hazardous Substances." December, 1987.
U.S. Department of Transportation, "Emergency Response Guidebook." Research and Special Programs Administration, Washington, DC, 1993.

U.S. Department of Transportation, U.S. Coast Guard, Chemical Hazards Response Information System (CHRIS), "Condensed Guide to Chemical Hazards, Volume I." Washington, DC, November 1992.

U.S. Department of Transportation, U.S. Coast Guard, Chemical Hazards Response Information System (CHRIS), "Hazardous Chemical Data Manual, Volume II." Washington, DC, November 1992.

U.S. Department of Transportation, "Emergency Response Guidebook." Research and Special Programs Administration, Washington, DC, 1993.

10 Selected Research and Discussion Topics

10.1. Synopsis

As new commodities are introduced into commerce, additional research will be required to characterize hazards and establish protection criteria. New or modified building features may become available to address hazards as they become better understood. Future research may require revision of some of the information provided in these Guidelines. Therefore, warehouse management must be prepared to respond to new information to fulfill its responsibilities to its shareholders, employees, and the community. The following sections address some specific topics covered in this guideline requiring additional research and discussion.

10.2. Commodity Hazards and Fire Protection Systems

Research is ongoing in several areas of commodity classification and sprinkler system design. The properties and hazards of most chemicals are understood and described in the literature. However, the burning characteristics and related sprinkler system design criteria to achieve control, suppression, or extinguishment is lacking for some commodities. To develop this information fire test programs with the following commodities are planned or underway:

- Aerosol products in stretch-wrapped plastic
- Flammable liquids in steel and plastic drums
- Flammable liquids (water soluble alcohols) in small plastic containers

- Flammable liquids in intermediate bulk containers (IBC's)
- Organic peroxide formulations
- Oxidizing swimming pool chemicals in plastic containers

As noted in Chapter 2, fire hazard classification systems are not based on uniform test methodologies for all commodities. In some cases, such as aerosol products, the classification is based on a quantifiable, heat of combustion criteria. However, with most ordinary commodities, the current classification is based on a subjective determination. Additional research to develop objective test methodologies for all commodities, including packaging systems, is needed.

Further research should be conducted toward establishing a better means to categorize the true fire hazard of all flammable and combustible liquids. The flash point, and in some cases boiling point, are measured values that are used for the current classification system. Additional properties, such as viscosity, dissolved combustible solids, and heat of combustion or heat release rate data should be included in a more comprehensive system.

Chapter 2 addresses chemical incompatibility, however most referenced information addresses binary combinations. A comprehensive database of chemical incompatibility, including multiple combinations, would be useful.

10.3. Design and Construction

Some building design issues should also be addressed through research and further discussion. For example, the use of roof-mounted heat and smoke vents, mechanical roof exhaust systems, and plastic skylight domes in combination with sprinkler systems are still being debated and has not been subjected to large-scale fire tests.

Design criteria for liquid containment and drainage systems also requires further research and discussion. Additionally, compatibility testing between floor coating materials and various chemical commodities would be useful. Proposed systems should be practical and cost effective.

APPENDIX A

Summary of NFPA 704 Marking System

HAZARD SCALE	HEALTH BLUE	FLAMMABILITY RED	REACTIVITY (Stability) YELLOW
4 SEVERE	Materials that, under emergency conditions, can be lethal.	Materials that will rapidly or completely vaporize at atmospheric pressure and normal ambient temperature or which are readily dispersed in air and which will burn readily.	Materials that in themselves are readily capable of detonation or explosive decomposition or explosive reaction at normal temperatures and pressures. This includes materials that are sensitive to localized thermal or mechanical shock at normal temperatures and pressures. Materials that have an instantaneous power density (product of heat of reaction and reaction rate) at 482°F (250°C) of 1000 W/mL or greater.
3 SERIOUS	Materials that, under emergency conditions, can cause serious or permanent injury.	Liquids and solids that can be ignited under almost all ambient temperature conditions. Materials in this degree produce hazardous atmospheres with air under almost all ambient temperatures or, though unaffected by ambient temperatures, are readily ignited under almost all conditions.	Materials that in themselves are capable of detonation or explosive reaction, but that require strong initiating source or that must be heated under confinement before initiation. This includes: • Materials that have an instantaneous power density (product of heat of reaction and reaction rate) at 482°F (250°C) at or above 100 W/mL and below 1000 W/mL; • Materials that are sensitive to thermal or mechanical shock at elevated temperatures and pressures; • Materials that react explosively with water without requiring heat or confinement.

153

Appendix A: Summary of NFPA 704 Marking System

HAZARD SCALE	HEALTH BLUE	FLAMMABILITY RED	REACTIVITY (Stability) YELLOW
2 MODERATE	Materials that, under emergency conditions, can cause temporary incapacitation or residual injury.	Materials that must be moderately heated or exposed to relatively high ambient temperatures before ignition can occur. Materials in this degree would not under normal conditions form hazardous atmospheres with air, but under moderate heating might release vapor in sufficient quantities to produce hazardous atmospheres with air.	Materials that readily undergo violent chemical change at elevated temperatures and pressures. This includes: Materials that have an instantaneous power density (product of heat of reaction and reaction rate) at 482°F (250°C) at or above 10 W/mL and below 100 W/mL; Materials that react violently with water or form potentially explosive mixtures with water.
1 SLIGHT	Materials that, under emergency conditions, can cause significant irritation.	Materials that must be preheated before ignition can occur. Materials in this degree require considerable preheating, under all ambient temperature conditions, before ignition and combustion can occur.	Materials that in themselves are normally stable, but that can become unstable at elevated temperatures and pressures. This includes: Materials that have an instantaneous power density (product of heat of reaction and reaction rate) at 482°F (250°C) at or above 0.01 W/mL and below 10 W/mL; Materials that react vigorously with water, but not violently; Materials that change or decompose on exposure to air, light, or moisture.
0 MINIMAL	Materials that, under emergency conditions, would offer no hazard beyond that of ordinary combustible materials.	Materials that will not burn.	Materials that in themselves are normally stable, even under fire conditions.

Source: NFPA 704 (See NFPA 704 for more complete information.)

APPENDIX

B | Summary of HMIS

HAZARD INDEX	HEALTH BLUE	FLAMMABILITY RED	REACTIVITY (Stability) YELLOW
4 SEVERE	Life-threatening, major or permanent damage may result from single or repeated exposures.	Very flammable gases or very volatile flammable liquids with flash points below 73°F (22.8°C) and boiling points below 100°F (37.8°C) (NFPA Class 1A).	Materials that are readily capable of detonation or explosive decomposition or reaction at normal temperatures and pressures.
3 SERIOUS	Major injury likely unless prompt action is taken and medical treatment is given.	Materials capable of ignition under almost all normal temperature conditions, including flammable liquids with flash points below 73°F (22.8°C) and boiling points above 100°F (37.8°C) as well as liquids with flash points between 73°F (22.8°C) and 100°F (37.8°C) (NFPA Class 1B and 1C).	Materials that are capable of detonation or explosive reaction but require a strong initiating source or must be heated under confinement before initiation; or materials that react explosively with water.
2 MODERATE	Temporary or minor injury may occur.	Materials that must be moderately heated before ignition will occur, including flammable liquids with flash points at or above 100°F (37.8°C) and below 200°F (93°C) (NFPA Class II and Class IIIA.)	Materials that, in themselves, are normally unstable and will readily undergo violent chemical change but will not detonate. These materials may also react violently with water.

155

Appendix B. Summary of HMIS

HAZARD INDEX	HEALTH BLUE	FLAMMABILITY RED	REACTIVITY (Stability) YELLOW
1 SLIGHT	Irritation or minor reversible injury possible.	Materials that must be preheated before ignition will occur. Flammable liquids in this category will have flash points (the lowest temperature at which ignition will occur) at or above 200°F (93°C) (NFPA Class IIIB).	Materials that are normally stable but can become unstable at high temperatures and pressures. These materials may react with water but they will not release energy violently.
0 MINIMAL	No significant risk to health.	Materials that are normally stable and will not burn unless heated.	Materials that are normally stable, even under fire conditions, and will not react with water.

PERSONAL PROTECTIVE INDEX	EQUIPMENT TYPE
A	Safety Glasses
B	Safety Glasses and Gloves
C	Safety Glasses, Gloves, and Synthetic Apron
D	Face Shield, Gloves, and Synthetic Apron
E	Safety Glasses, Gloves, and Dust Respirator
F	Safety Glasses, Gloves, Synthethic Apron, and Dust Respirator
G	Safety Glasses, Gloves, and Vapor Respirator
H	Splash Goggles, Gloves, Synthetic Apron, and Vapor Respirator
I	Safety Glasses, Gloves, and Combination Dust and Vapor Respirator
J	Splash Goggles, Gloves, Synthetic Apron, and Combination Dust and Vapor Respirator
K	Airline Hood or Mask, Gloves, Full Protective Suit, and Boots
X	Ask your supervisor for specialized handling directions.

Source: NPCA Hazardous Materials Identification System

APPENDIX C

United Nations and U.S. Department of Transportation Hazardous Materials Classes

Class 1	**Explosives**
Division 1.1	Explosives with a mass explosion hazard
Division 1.2	Explosives with a projection hazard
Division 1.3	Explosives with predominately a fire hazard
Division 1.4	Explosives with no significant blast hazard
Division 1.5	Very insensitive explosives; blasting agents
Division 1.6	Extremely insensitive detonating substances
Class 2	**Gases**
Division 2.1	Flammable gas
Division 2.2	Nonflammable, nonpoisonous compressed gas
Division 2.3	Gas poisonous by inhalation
Division 2.4	Corrosive gas (Canadian)
Class 3	**Flammable liquid and Combustible liquid**
	Flammable liquid—Flash point <141°F (60.5°C)
	Combustible liquid—Flash point ≥141°F (60.5°C)
Class 4	**Flammable solid; Spontaneously combustible material; and Dangerous when wet material**
Division 4.1	Flammable solid
Division 4.2	Spontaneously combustible material
Division 4.3	Dangerous when wet material
Class 5	**Oxidizers and Organic peroxides**
Division 5.1	Oxidizer
Division 5.2	Organic peroxide

Class 6   **Poisonous material and Infectious substance**
   Division 6.1   Poisonous materials
           • Poison—Packing Group I and II
           • Poison—Packing Group III
   Division 6.2   Infectious substance

Class 7   Radioactive material

Class 8   Corrosive material

Class 9   Miscellaneous hazardous material

APPENDIX

D Additional Resources

Additional information can be obtained from these sources:

American Heart Association
7272 Greenville Rd.
Dallas, TX 75231

American National Standards Institute
11 W. 42nd St., Floor 13
New York, NY 10036

American Red Cross
8111 Gatehouse Road
Falls Church, VA 22042

American Society for Testing and Materials
1916 Race Street
Philadelphia, PA 19103-1187

Center for Chemical Process Safety
American Institute of Chemical Engineers
345 East 47th Street
New York, NY 10017

Chemical Manufacturers Association
1300 Wilson Boulevard
Arlington, VA 22209

Factory Mutual System
1151 Boston-Providence Turnpike
Norwood, MA 02062

Industrial Risk Insurers
85 Woodland Street
Hartford, CT 06105-1226

Insurance Institute of America
720 Providence Road
Malvern, PA 19355-0716

MidWest Plan Service
Agricultural and Biosystems Engineering Department
122 Davidson Hall
Iowa State University
Ames, IA 50011-3080

National Fire Protection Association
1 Battery March Park
Quincy, MA 02269

National Safety Council
1121 Spring Lake Drive
Itasca, IL 60143

Department of Transportation
United States Coast Guard
2100 Second Street, SW
Washington, D.C. 20593-0001

Glossary of Terms

Auto-Ignition Temperature: The minimum temperature required to initiate or cause self-sustained combustion of a substance, in air, independent of any heated or heating element.

Base Flood: See Special Hazard Flood Area

Biological Oxygen Demand: The amount of dissolved oxygen in water, given in lbs. (kgs) or % that is consumed by biological oxidation of a chemical.

CAS Registry Number: A unique number having up to nine digits that is assigned to a chemical by the Chemical Abstracts Service of the American Chemical Society. With the exception of aqueous solutions, mixtures are not covered under this registry.

Catalyst: A chemical substance that accelerates the rate of a chemical reaction by lowering the energy of activation required for the chemical reaction to occur.

Combustible Liquid: A liquid that has a closed-cup flash point at or above 100°F (37.8°C), as determined by test. Combustible liquids are further classified as follows:
- Class II—A liquid that has a flash point at or above 100°F (37.8°C) but below 140°F (60°C).
- Class IIIA—A liquid that has a flash point at or above 140°F (60°C) but below 200°F (93°C).
- Class IIIB—A liquid that has a flash point at or above 200°F (93°C).

Deflagration: Propagation of a reaction zone into the unreacted material at a speed that is less than the speed of sound in the unreacted material. Where a blast wave is produced that has the potential to cause damage, the term *explosive deflagration* may be used.

Detonation: Propagation of a reaction zone into the unreacted material at a speed that is equal to or greater than the speed of sound in the unreacted material.

Endothermic: A physical or chemical change that requires or is accompanied by the absorption of heat.

Endothermic Chemical Reaction: A reaction involving one or more chemicals resulting in one or more new chemical species and the absorption of heat.

Exothermic: A physical or chemical change accompanied by the evolution of heat.

Exothermic Chemical Reaction: A reaction involving one or more chemicals resulting in one or more new chemical species and the evolution of heat.

Fire Point: The minimum temperature at which a flammable or combustible liquid, as herein defined, and some volatile combustible solids will evolve sufficient vapor to produce a mixture with air that will support sustained combustion when exposed to a source of ignition, such as a spark or flame.

Flammable Gas (NFPA 55): A gas that is flammable in a mixture of 13 percent or less (by volume) with air, or the flammable range with air is wider than 12 percent regardless of the lower limit, at atmospheric temperature and pressure.

Flammable Liquid: A liquid that has a closed-cup flash point below 100°F (37.8°C) as determined by test. Flammable liquids are further classified as follows:
- Class IA—A liquid that has a flash point below 73°F (22.8°C) and boiling point below 100°F (37.8°C).
- Class IB—A liquid that has a flash point below 73°F (22.8°C) and a boiling point at or above 100°F (37.8°C).
- Class IC—A liquid that has a flash point at or above 73°F (22.8°C), but below 100°F (37.8°C).

Flammable Range: Those concentrations of a combustible gas or vapor in air, measured as volume percent in air, at which flaming combustion can occur. The flammable range has an upper and lower limit. The flammable range is bounded by an upper flammable limit (UFL) and a lower flammable limit (LFL).

Flash Point: The minimum temperature at which a flammable or combustible liquid, as herein defined, and some volatile combustible solids will evolve sufficient vapor to produce an ignitable mixture with air near the surface of the liquid or solid or within the test vessel used.

Glossary of Terms 163

Unlike a fire point, ignition at the flash point temperature may not result in sustained combustion. There are several flash point test methods, and flash points may vary for the same material depending on the method used. Consequently, it is important that the test method is indicated when the flash point is given (150°PMCC, 200°TCC, etc.,). A closed cup type test is used most frequently for regulatory purposes. Flash point test methods are as follows:

Cleveland Open-Cup (CC)

Pensky Martens Closed-Cup (PMCC)

Setaflash Closed Tester (SETA)

Tagliabue (Tag) Closed-Cup (TCC)

Tagliabue (Tag) Open-Cup (TOC)

Hazardous Waste (40 CFR 261):
Hazardous waste can be classified as follows:
- *Ignitable*—Based upon the flash point of a liquid waste; for a solid, the capability to cause fire through friction or absorption of moisture, and to burn vigorously and persistently; solids that meet the 49 CFR definition of oxidizer; and, compressed gases that are ignitable under the DOT definition.
- *Corrosive*—Liquid wastes that have a pH of *plain* 2 or 12.5, or that corrode steel at a rate of greater than 0.25 inch per year.
- *Reactive*—Wastes that are unstable and readily undergo violent change; that react violently with water or when mixed with water generate toxic vapors or fumes; that are cyanide or sulfide bearing and can generate toxic gases, vapors, or fumes at pH conditions between 2 and 12.5; that are readily capable of detonation or explosion at standard temperature and pressure if subjected to a strong initiating force or if heated under confinement; or DOT forbidden explosives.
- *Toxic*—Liquid wastes or extract from waste solids that fail the Toxicity Characteristic Leaching Procedure (TCLP) analytical test because they contain certain designated metals, pesticides, or organic chemicals at concentrations equal to or, in excess of, specified regulatory limits.

Inhibitor: A chemical substance used to prevent or stop a chemical reaction, such as polymerization, from occurring.

IUPAC Name: A chemical name derived from a formal system of nomenclature employing a fundamental principle that each specific compound will have a different name. The system was developed and is maintained by the International Union of Pure and Applied Chemistry.

Lacustrine Flood: A flood resulting from an overflow of a lake or pond.

LC$_{50}$: Lethal Concentration, due to respiratory exposure, at which approximately 50% of exposed test animals will die.

LD$_{50}$: Lethal Dosage, due to ingestion or skin absorption, at which approximately 50% of exposed test animals will die.

Nonflammable Gas (NFPA 55): A gas that does not meet the definition of a flammable gas.

Ordinary Commodity: A term used to describe a commodity not having special properties or hazards and is categorized by its relative fire hazard. Ordinary commodities can be further classified as follows:
- Noncombustible (NFPA 13)—A material that, in the form in which it is used and under the conditions anticipated, will not ignite, burn, support combustion, or release flammable vapors when subjected to fire or heat. Materials that are reported as passing ASTM E136, *Standard Test Method for Behavior of Materials in a Vertical Tube Furnace at 750°C*, shall be considered noncombustible materials.
- Class I (NFPA 231/231C)—Class I commodity is defined as essentially noncombustible products on wood pallets, in ordinary corrugated cartons with or without single-thickness dividers, or in ordinary paper wrappings, all on wood pallets. Such products are permitted to have a negligible amount of plastic trim, such as knobs or handles.
- Class II (NFPA 231/231C)—Class II commodity is defined as Class I products in slatted wooden crates, solid wooden boxes, multi-wall corrugated cartons, or equivalent combustible packaging material on wood pallets.
- Class III (NFPA 231/231C)—Class III commodity is defined as wood, paper, natural fiber cloth, or Group C plastics or products thereof; on wood pallets. Products are permitted to contain a limited amount of Group A or B plastics. Wood dressers with plastic drawer glides, handles, and trim are examples of a commodity with a limited amount of plastic.
- Class IV (NFPA 231/231C)—Class IV commodity is defined as Class I, II, or III products containing an appreciable amount of Group A plastics in a paperboard carton or Class I, II, or III products with Group A plastic packing in paperboard cartons on wood pallets. Group B plastics and free-flowing Group A plastics are also included in this class.

Examples of Class IV products are: small appliances, typewriters, and cameras with plastic parts; plastic-backed tapes and synthetic fabrics or clothing. An example of packing material is a metal product in a foamed plastic cocoon in a corrugated carton.
- Classification of Plastics, Elastomers, and Rubber (NFPA 231C) NOTE: The following categories are based on unmodified plastic materials. The use of fire or flame-retarding modifiers or the physical form of the material may change the classification.

Group A
- ✓ ABS (acrylonitrile–butadiene–styrene copolymer)
- ✓ Acetal (polyformaldehyde)
- ✓ Acrylic (polymethyl methacrylate)
- ✓ Butyl rubber
- ✓ EPDM (ethylene–propylene rubber)
- ✓ FRP (fiberglass reinforced polyester)
- ✓ Natural rubber
- ✓ Nitrile rubber (acrylonitrile–butadiene rubber)
- ✓ PET (polyethylene terephthalate)
- ✓ Polybutadiene
- ✓ Polycarbonate
- ✓ Polyester elastomer
- ✓ Polyethylene
- ✓ Polypropylene
- ✓ Polystyrene
- ✓ Polyurethane
- ✓ PVC (polyvinyl chloride—highly plasticized, e.g., coated fabric, unsupported film)
- ✓ SAN (styrene acrylonitrile)
- ✓ SBR (styrene–butadiene rubber)

Group B
- ✓ Cellulosics (cellulose acetate, cellulose acetate butrate, ethyl cellulose)
- ✓ Chloroprene rubber
- ✓ Fluoroplastics (ECTFE: ethylene–chlorotrifluoro–ethylene copolymer; ETFE: ethylene–tetrafluoroethylene copolymer; FEP: fluorinated ethylene–propylene copolymer)
- ✓ Nylon (nylon 6, nylon 6/6)
- ✓ Silicone rubber

Group C
- ✓ Fluoroplastics (PCTFE: polychlorotrifluoroethylene; PTFE: polytetrafluoroethylene)

✓ Melamine (melamine formaldehyde)
✓ Phenolic
✓ PVC (polyvinyl chloride—rigid or lightly plasticized, e.g., pipe, pipe fittings)
✓ PVDC (polyvinylidene chloride)
✓ PVF (polyvinyl fluoride)
✓ PVDF (polyvinylidene fluoride)
✓ Urea (urea formaldehyde)

Organic Peroxide (NFPA 43B): Any organic compound having a double oxygen or "peroxy" (–O–O–) group in its chemical structure.

Organic Peroxide Formulation (NFPA 43B): A pure organic peroxide or a mixture of one or more organic peroxides with one or more other materials in various combinations and concentrations. Organic Peroxide Formulations can be further classified as follows:
- Class I—Those formulations that are capable of deflagration but not detonation.
- Class II—Those formulations that burn very rapidly and that present a severe reactivity hazard.
- Class III—Those formulations that burn very rapidly and that present a moderate reactivity hazard.
- Class IV—Those formulations that burn in the same manner as ordinary combustibles and that present a minimal reactivity hazard.
- Class V—Those formulations that burn with less intensity than ordinary combustibles or do not sustain combustion and that present no reactivity hazard.

Oxidizing Gas (NFPA 55): A gas that can support and accelerate combustion of other materials.

Oxidizer (NFPA 430): Any material that readily yields oxygen or other oxidizing gas, or that readily reacts to promote or initiate combustion of combustible materials. Examples of other oxidizing gases include Bromine, Chlorine, and Fluorine. Oxidizers can be further classified as follows:
- Class 1—An oxidizer whose primary hazard is that it slightly increases the burning rate but does not cause spontaneous ignition when it comes in contact with combustible materials.
- Class 2—An oxidizer that will cause a moderate increase in the burning rate or that causes spontaneous ignition of combustible materials with which it comes in contact.
- Class 3—An oxidizer that will cause a severe increase in the burning rate of combustible materials with which it comes in contact or

that will undergo vigorous self-sustained decomposition due to contamination or exposure to heat.
- Class 4—An oxidizer that can undergo an explosive reaction due to contamination or exposure to thermal or physical shock. In addition, the oxidizer will enhance the burning rate and may cause spontaneous ignition of combustibles.

Polymerization: A chemical reaction in which one or more small molecules combine to form larger molecules. A hazardous polymerization is such a reaction which takes place at a rate which releases large amount of energy.

Pyrophoric: Term used to describe a substance capable of spontaneous combustion when in contact with air.

Pyrophoric Gas (NFPA 55): A gas that will spontaneously ignite in air at or below a temperature of 130°F (54°C).

Registry of Toxic Effects of Chemical Substances: A system developed by the National Institute for Occupational Safety and Health in which a nine position alphanumeric designation is assigned to a chemical name.

Riverine Flood: A flood resulting from an overflow of a river, stream or creek..

Self-Accelerating Decomposition Temperature: The minimum temperature that a mass of material, capable of an exothermic decomposition reaction, must be held such that the heat of decomposition exceeds the amount of energy lost to the surroundings. This will result in an increase in the mass temperature and acceleration of the decomposition reaction rate.

Special Hazard Flood Area: An area subject to inundation by a flood having a one-percent or greater probability of being equaled or exceeded during any given year. Also known as the 100-year flood or base flood.

Specific Gravity: A dimensionless, temperature dependent ratio of the density of one substance with that of a reference substance. For solids and liquids the reference substance is water at 39°F (4°C). For gases and vapors, the reference substance is dry air at 60°F (15.6°C).

Stoichiometric Concentration: A term used to describe a balanced ratio of chemical reactants that would result in all of the chemical reactants becoming products if a chemical reaction were to occur.

Toxic Gas (NFPA 55): A gas having a health hazard rating of 3 or 4 as defined in NFPA 704.

TL$_M$: Median Tolerance Limit at which approximately 50% of the exposed species will show abnormal behavior including death under the conditions of concentration and time given.

UN/NA Number: A four digit number used by both the United Nations and the U.S. Department of Transportation. This number is assigned to a hazardous chemical or group of hazardous chemicals.

Vapor–Air Density: The weight of a volume of pure gas or vapor compared to the weight of an equal volume of dry air at the same temperature and pressure.

Vapor Pressure: The pressure exerted at a given temperature above a liquid or solid.

Index

Accident scenarios, 29
Aerosols, 12–13, 90, 151–152
Airborne effluents, 86
Air monitoring/sampling, 49, 51
Arctic freeze, 62

Building codes, 67–70
Building construction
   combustible, 86–88
   damage limiting, 103–104
   fire resistive, 86–88
   noncombustible, 86–88
   wind resistant, 107

Changes in occupancy, 32
Chemical identity, 3–4
Chemical properties, 6
Chemical protective clothing, 41–49
Chemical releases, 144–146
Chemical resistance, 41–42
CMA responsible care, 148
Coatings and sealers, 84–85
Commodity compatibility and
   separation, 18–20
Commodity classification, 5, 151–152
   reference list, 11
Commodity hazard, 3
Concrete, 82–85
Container and packaging systems, 17

Decontamination procedure, 51
Deflagration, 99–104
Department of Transportation, 17, 157–158
Detection alarms, 37

Earthquakes, 58, 105–106
Egress design, 70–72
Emergency notification procedures, 139
Emergency plan, 135–149
Emergency responders, 64, 138
Emergency spill response, 49–52, 140–146
Employees, 36–37, 39–40, 136
Engineering controls, 37
Environmental assessments, 56
Environmental protection
   agency, 11–12, 147
Evacuation, 70–72, 136, 139

Federal Emergency Management
   Agency, 59
Fire doors, 93–97
Fire control, 51, 115–119
   environmental hazards, 10
   fire/explosion hazards, 7
   human health hazards, 8
Hazardous use facilities, 69

169

Hazardous wastes, 12
High risk materials, 29
Hurricanes, 59–61, 107

Ignition sources, 26–27, 102, 113
Incident response, 140
Incompatible chemicals, 19–20
Inspection and test programs, 129–133
International code council 68–69
Inventory identification and
  management systems, 29–31

Labels, 38
Life safety code, 68, 71–73
Lightning, 62, 107
Loss scenarios, 135, 138

Maintenance programs, 131–133
Material safety data sheets, 38–39

National Fire Protection Association,
  12–16, 68, 153, 154
National Paint and Coatings
  Association, 16–17, 155–156
Natural peril exposures, 57–62
Natural peril mitigation, 104–108

Ordinary commodities, 14
Organic peroxide formulations, 14–15,
  99, 152
OSHA, 44, 147–148

Oxidizers, 15–16, 100, 152

Packing groups, 18
Personal protective equipment 40–49
Personal protective index 16
Pesticides, 11
Pipelines, 63
Public relations, 140
Public utilities, 64
RCRA, 148
Remediation, 140
Respirators, 44–47
Riot/civil commotion, 63
Risk management plan, 135–149
  cut-off, 20–21
  detached, 20–21
  outdoor, 108–109
  segregated, 20–21

Storage use facilities, 69

Tornadoes, 60–62, 107
Training programs, 36, 39–40.46–48
Travel distance, 72–73

United Nations, 17–18, 157–158

Ventilation, 97–99, 102

Windstorm mitigation, 107–108

Publications Available from the
CENTER FOR CHEMICAL PROCESS SAFETY
of the
AMERICAN INSTITUTE OF CHEMICAL ENGINEERS

CCPS Guidelines Series

Guidelines for Safe Warehousing of Chemicals
Guidelines for Postrelease Mitigation in the Chemical Process Industry
Guidelines for Integrating Process Safety Management, Environment, Safety, Health, and Quality
Guidelines for Use of Vapor Cloud Dispersion Models, Second Edition
Guidelines for Evaluating Process Plant Buildings for External Explosions and Fires
Guidelines for Writing Effective Operations and Maintenance Procedures
Guidelines for Chemical Transportation Risk Analysis
Guidelines for Safe Storage and Handling of Reactive Materials
Guidelines for Technical Planning for On--Site Emergencies
Guidelines for Process Safety Documentation
Guidelines for Safe Process Operations and Maintenance
Guidelines for Process Safety Fundamentals in General Plant Operations
Guidelines for Chemical Reactivity Evaluation and Application to Process Design
Tools for Making Acute Risk Decisions with Chemical Process Safety Applications
Guidelines for Preventing Human Error in Process Safety
Guidelines for Evaluating the Characteristics of Vapor Cloud Explosions, Flash Fires, and BLEVEs
Guidelines for Implementing Process Safety Management Systems Guidelines for Safe Automation of Chemical Processes
Guidelines for Engineering Design for Process Safety
Guidelines for Auditing Process Safety Management Systems
Guidelines for Investigating Chemical Process Incidents
Guidelines for Hazard Evaluation Procedures, Second Edition with Worked Examples
Plant Guidelines for Technical Management of Chemical Process Safety, Revised Edition
Guidelines for Technical Management of Chemical Process Safety

Guidelines for Chemical Process Quantitative Risk Analysis
Guidelines for Process Equipment Reliability Data with Data Tables
Guidelines for Safe Storage and Handling of High 'Toxic Hazard Materials
Guidelines for Vapor Release Mitigation

CCPS Concepts Series

Inherently Safer Chemical Processes. A Life--Cycle Approach
Contractor and Client Relations to Assure Process Safety
Understanding Atmospheric Dispersion of Accidental Releases
Expert Systems in Process Safety
Concentration Fluctuations and Averaging Time in Vapor Clouds

Proceedings and Other Publications

Proceedings of the International Conference and Workshop on Risk Analysis in Process Safety, 1997
Proceedings of the International Conference and Workshop on Process Safety Management and Inherently Safer Processes, 1996
Proceedings of the International Conference and Workshop on Modeling and Mitigating the Consequences of Accidental Releases of Hazardous Materials, 1995.
Proceedings of the International Symposium and Workshop on Safe Chemical Process Automation, 1994
Proceedings of the International Process Safety Management Conference and Workshop, 1993
Proceedings of the International Conference on Hazard Identification and Risk
Analysis, Human Factors, and Human Reliability in Process Safety, 1992
Proceedings of the International Conference and Workshop on Modeling and Mitigating the Consequences of Accidental Releases of Hazardous Materials, 1991.
Safety, Health and Loss Prevention in Chemical Processes: Problems for Undergraduate Engineering Curricula